China's Telecommunications Revolution

China's Telecommunications Revolution

Eric Harwit

OXFORD
UNIVERSITY PRESS

Great Clarendon Street, Oxford OX2 6DP

Oxford University Press is a department of the University of Oxford.
It furthers the University's objective of excellence in research, scholarship,
and education by publishing worldwide in

Oxford New York

Auckland Cape Town Dar es Salaam Hong Kong Karachi
Kuala Lumpur Madrid Melbourne Mexico City Nairobi
New Delhi Shanghai Taipei Toronto

With offices in

Argentina Austria Brazil Chile Czech Republic France Greece
Guatemala Hungary Italy Japan Poland Portugal Singapore
South Korea Switzerland Thailand Turkey Ukraine Vietnam

Oxford is a registered trade mark of Oxford University Press
in the UK and in certain other countries

Published in the United States
by Oxford University Press Inc., New York

© Eric Harwit 2008

The moral rights of the author have been asserted
Database right Oxford University Press (maker)

First published 2008

All rights reserved. No part of this publication may be reproduced,
stored in a retrieval system, or transmitted, in any form or by any means,
without the prior permission in writing of Oxford University Press,
or as expressly permitted by law, or under terms agreed with the appropriate
reprographics rights organization. Enquiries concerning reproduction
outside the scope of the above should be sent to the Rights Department,
Oxford University Press, at the address above

You must not circulate this book in any other binding or cover
and you must impose the same condition on any acquirer

British Library Cataloguing in Publication Data

Data available

Library of Congress Cataloging in Publication Data

Data available

Typeset by SPI Publisher Services, Pondicherry, India
Printed in Great Britain
on acid-free paper by
Biddles Ltd., King's Lynn, Norfolk

ISBN 978–0–19–923374–8

1 3 5 7 9 10 8 6 4 2

To Hiroko and Mari

Preface

The speed of China's ascent to economic superpower status has few parallels in history. When I began to write this book, I was wondering how this astonishing development had come about. The rise of the telecommunications industry in the People's Republic seemed to provide a mirror to developments elsewhere in the country. Looking at this industry, I wondered how the particular mix of a centrally governed society venturing into free market enterprise had managed to succeed in a field in which foreign powers appeared to have far more experience.

Did China simply follow the historic developments pioneered by other countries, or did it chart its own destiny? What was the role that central government took on? How did diverse government ministries and regional officials imprint their own preferences on developments? In a society as tightly controlled as China's, how did the country's vast rural regions fare in the telecommunications revolution, and how did the central government keep tabs on political thoughts expressed across the rapidly growing Internet?

Chinese enterprises seemed to have attracted foreign investments in part to help the new home-grown ventures achieve rapid success. How did the government assure that China would retain ultimate control, and keep most of the created wealth at home? Now that China has joined the World Trade Organization, some of the government's past interventions may have to be modified. What, if any, are the changes one can begin to discern?

These questions, and answers to them, have occupied my research for the past decade. I have tried in this book to express my findings as clearly as possible, but China's telecommunications revolution is still ongoing, and much may still change in the years ahead. The book should therefore be seen as a work in progress, reporting on the ways China's telecommunications industry has evolved, as a harbinger of all that still lies ahead.

E.H.

Acknowledgments

I extend thanks to the many scholars, students, and business people in the telecommunications field who offered their help and advice over the years it took me to compile this book.

My conversations and collaborations with Duncan Clark, the Chairman of BDA (China), were integral and invaluable for framing many of the book's arguments. In 2001, I published an earlier version of Chapter 4 with Mr Clark, and have continued to value his friendship over the years.

I have also enjoyed talking with other telecommunications specialists. Craig Watts, who has held a number of important positions in the field in China, was a source of inspiration for several of the book's points, and I appreciate his taking time to discuss and debate the issues of Chinese government involvement in the telecommunications sector. Kevin Bunka was also a ready source of ideas and knowledge on the evolution of the field, and I value his input to my understanding. J. P. Huang, a successful businessman in many areas and a valued friend in China, helped arrange several key interviews during my years of fieldwork. Bu Wei also was generous with her time and resources at the Chinese Academy of Social Sciences. Many other Chinese business people and scholars with whom I discussed key issues of telecommunications and the Internet chose to remain anonymous. However, they were also integral to the development of my work.

Fellow American scholars of China's political economy and scientific development were also helpful in providing their opinions on various parts of my work. Among these, Denis Simon and Richard P. Suttmeier were most encouraging.

Students both at the University of Hawaii and in China have been extremely helpful in finding supporting data for my work. Among these, Wang Qinghong in Hawaii consistently helped with key Chinese documents. In China, Peking University students Zhang Yue, Song Qian, and Su Li conducted database and library assistance. Other students at the University of Hawaii who provided research assistance included Allison

Acknowledgments

Pan, Michelle Valle, Chris Gin, and Atsushi Ouchi. Students in several of my graduate course seminars, in particular one that focused on telecommunications in East Asia, were also helpful as I presented my ideas in class, and learned from their own research efforts.

My research during the past ten years was supported by several sources of funding. Among grants I received were a postdoctoral fellowship from the University of California; a Fulbright Program research grant; and funds from the University of Hawaii's Center for International Business Education and Research (CIBER) and National Resource Center for East Asia (NRC-EA). During a 2000–1 sabbatical leave spent partially at Stanford University, I appreciated the hosting of the Asia Pacific Research Center (APARC). I spent several summers conducting research at Peking University using the University of Hawaii's scholar exchange program.

Finally, members of my family were also key contributors to this project. My father, Martin Harwit, read each of my chapters and offered encouragement at every step, as did my mother, Marianne Harwit. More patient, however, were my wife Hiroko and daughter Mari, who generously accommodated themselves to my years of work on this project. I could not have finished this work without their support and encouragement.

Contents

List of Figures	xii
List of Tables	xiii
List of Abbreviations	xv
1. Industrial Policy and the Chinese Telecommunications Industry	1
2. China's Telecommunications History and Policy Trajectory	18
3. Telecommunications Competition in the 1990s and 2000s	43
4. China's Internet and Government Policy	79
5. Building the Network: The Role of Telecommunications Equipment Companies	112
6. Telecommunications in Shanghai: A Case Focus on the Municipal Government's Role	135
7. The Digital Divide of Telephones and the Internet	158
8. Industrial Policy and Lessons of the Telecommunications Revolution	183
Notes	190
Bibliography	219
Index	231

List of Figures

2.1.	Number of telephone subscribers, 1949–1975	31
2.2.	Number of telephone subscribers, 1975–2007 (selected years)	39
3.1.	Major players in telecommunications services, 1949–2007	51
4.1.	Evolution of main academic and commercial interconnecting data networks and their regulators, 1987–2007 (from date operations began)	84
4.2.	Growth in number of Internet users, year end 1996–2006	89
4.3.	Control hierarchy and revenue flows of Internet service provision under the MII	91
4.4.	Numbers of .cn registered domain names, year-end 2000–2006	101
4.5.	Internet use by age, end 2006	106
4.6.	Changing Internet use by gender, selected years, 1998–2006	107
5.1.	Central office switch capacity (in total number of phone ports), 1949–1990	115
5.2.	Number of fixed-line telephone subscribers, 1980–2006	122
5.3.	Growth in number of mobile phone subscribers and Internet users, 1995–2006	130
6.1.	Number of Shanghai fixed-line and mobile telephone subscribers and Internet users, 1985–2006	140
6.2.	Shanghai's national–local telecommunications hierarchy, with key corporations	141
7.1.	Internet penetration, percentage by region, end 2006	171
7.2.	Status of the telecommunications revolution in elite urban, urban, and rural areas	173

List of Tables

2.1. Fixed-asset investment in telecommunications as percent of gross domestic product (GDP), 1978–2006, (selected years)	38
2.2. Numbers of telephones and teledensity for China and other selected nations, 1999 and 2005	39
3.1. Unicom's shareholders and their ownership percentage	48
3.2. Mobile phone customer growth and market share, 1994–1997, for Unicom and China Telecom, at year end	52
3.3. Partial list of Unicom's regional shareholders	55
3.4. Examples of Unicom 'CCF' joint ventures with foreign firms	56
3.5. Mobile phone subscriber growth and market share, 1997–2007, selected years, for Unicom and China Telecom / China Mobile, at year end (except as noted)	65
3.6. Monthly price basket for fixed-line and mobile telephone use for China and selected Asian nations, 2004	76
4.1. Major networks and their leased international bandwidth (in Megabits per second) (selected years, 1998–2007)	88
4.2. Internet use and penetration for China and selected nations, 2005	90
4.3. Monthly price basket for Internet use for China and selected Asian nations, 2004	94
4.4. Top web domain sites for all users, February 2001 and June 2007	102
4.5. Comparison of selected activities of American and Chinese Internet use patterns	108
5.1. Central office switch sales, selected years, 1990–2003, in millions of ports	119
7.1. Number of fixed-line urban and rural subscribers, selected years, 1951–2006	163
7.2. Rural telephone access in selected areas of highest and lowest penetration, 2001 and 2005 (based on 2001 ranking of top 9 and bottom 5)	166

Tables

7.3. Percentage of total population using the Internet in areas of high and low use, 2000–2006, grouped according to top 5 and bottom 5 ranking areas in the 2006 count … 170

7.4. Average per capita monthly income for areas with high and low Internet use, in 2006 … 172

List of Abbreviations

ARPA	Advanced Research Projects Agency
AT&T	American Telephone and Telegraph Company
B2B	business to business
B2C	business to consumer
BISC	Beijing International Switching Company
BT	British Telecom
C2C	consumer to consumer
CAS	Chinese Academy of Sciences
CCF	Chinese–Chinese–Foreign
CDMA	code division multiple access
CERNET	China Education and Research Network
CITIC	China International Trust and Investment Corporation
CSNet	China Satellite Network
CSTNet	China Science and Technology Network
DGT	Directorate General of Telecommunications
FCC	Federal Communications Commission
FT	France Telecom
GDP	gross domestic product
GNI	Gross National Income
GPRS	general packet radio service
GSM	Global System for Mobile
ICC	Interstate Commerce Commission
ICP	Internet content provider
ICT	information and communication technology
IHEP	Institute of High Energy Physics
IP	Internet protocol
IPO	initial public offering

Abbreviations

IPTV	Internet protocol television
ISP	Internet service provider
ITA	Imperial Telegraph Administration
KDD	Kokusai Denshin Denwa
KMT	Kuomintang
KPTC	Kenya Posts and Telecommunications Corporation
MAI	Ministry of Aerospace Industry
Mbps	Megabits-per-second
MEI	Ministry of Electronics Industry
MII	Ministry of Information Industry
MITI	Ministry of International Trade and Industry
MOP	Ministry of Electric Power
MOR	Ministry of Railways
MOU	memorandum of understanding
MPT	Ministry of Posts and Telecommunications
NEC	Nippon Electric Company
NSF	National Science Foundation
NTT	Nippon Telegraph and Telephone Corporation
OECD	Organization for Economic Cooperation and Development
P&T	Posts and Telecommunications
PHS	personal handy system
PLA	People's Liberation Army
PRC	People's Republic of China
PTA	Posts and Telecommunications Administration
PTA	provincial telecommunications administration
PTIC	China Posts and Telecommunications Industry Corporation
REA	Rural Electrification Administration
SAIL	Shanghai Alliance Investment Company Ltd.
SARFT	State Administration for Radio, Film, and Television
SASAC	State-Owned Assets Supervision and Administration Commission
SCN	Shanghai Cable Network
SDPC	State Development and Planning Commission
SEC	State Economic Commission
SEC	State Education Commission

Abbreviations

SETC	State Economic and Trade Commission
SII	Shanghai Information Investment Corporation
SMS	Short-message services
SOFTEC	Shanghai Optical Fiber Telecommunications Engineering Corporation
SPC	State Planning Commission
SPTA	Shanghai Posts and Telecommunications Administration
SSTIC	Shanghai Science and Technology Investment Corporation
TD-SCDMA	Time Division Synchronous CDMA
TFP	total factor productivity
UNESCO	United Nations Educational, Scientific, and Cultural Organization
USF	Universal Service Fund
VOIP	voice over Internet protocol
WTO	World Trade Organization
ZTA	Zone de télécommunications avancée
ZTE	Zhongxing Telecommunications Equipment Company

1
Industrial Policy and the Chinese Telecommunications Industry

As a young exchange student on my first trip to China in 1982, I soon found many of the conveniences I had come to take for granted in the United States were absent at my college campus in Beijing. Supplies of clean drinking water were several minutes walk away, showers were only available a few times a week, and frigid buildings had nearly non-existent heating during the harsh winter.

One other 'necessity' I soon learned to do without was a telephone. At the time, telephones in private Chinese homes were virtually unheard of, and office phones were off limits except for official use. To make the occasional long-distance call home, I had to bicycle for about an hour to the central telephone exchange office in downtown Beijing, and wait perhaps an hour for the call to go through.

Since my first stay in China, telecommunications conditions improved year by year, and foreign students in the new century, at least in urban areas, face few of the problems I did more than two decades ago. The number of telephone subscribers in the People's Republic of China (PRC) soared from only 2 million in 1980 to more than 360 million by 2007, and mobile phone use leaped from 4 million subscribers in 1995 to some 500 million. In both categories, China ranked number one in the world. Internet users as of mid-2007 numbered more than 160 million, and the PRC expected to pass the US in total numbers of connected citizens within the next few years.

China has arguably been the most successful developing nation in history for spreading telecommunications access in a remarkably short period of time. Yet few analysts have taken on the challenge of explaining how central government policies, local government actions,

Industrial Policy

and orchestrated foreign investment played a key role in building this massive communications network.

The main argument of this book is that government control and ownership of telecommunications operations was key to the sector's growth in the critical 1980s and 1990s. Revenue from corporate and citizen use of the network funded expansion of the basic telecommunications infrastructure. Top decision makers in the government's State Council as well as in the telecommunications bureaucracy skillfully employed industrial policies to guide the network's growth. Foreign corporations were prohibited from managing telephone and Internet networks so that the government could reap the lucrative revenue rewards, but overseas companies did play a major role in providing equipment to link the network hardware. City governments played a supporting yet vital role by channeling funds to local corporations overseeing construction of the industry.

Some of the results of the Chinese government effort were immediately obvious, as citizens came to enjoy vast new opportunities for communication. One of the goals of this book is to describe how these fundamental changes were put in place, and the economic and social impact they have had on nearly a fifth of the world's population.

By the new decade of the 2000s, new issues in the sector portended forthcoming changes in the policy path. The challenge of global forces dictated that the protectionist industrial policies had begun to run their course. Citizens in wealthy parts of the nation began to demand a new generation of telecommunications equipment and more efficient services, and the country's entry to the World Trade Organization (WTO) portended a renewed challenge by foreign telecommunications corporations of all kinds. At the same time, the PRC faced a further hurdle, that of extending basic telecommunications services to those in the poorer inland parts of the nation. Finally, the potentially explosive social and political challenges of widespread Internet use and the rapid spread of knowledge made some leaders, concerned about information control, wonder whether expansion of the network had come at too great a price. The text analyzes the ways central and local government policies shifted at the dawn of the new decade, and how foreign investors have reacted to new opportunities in a post-WTO China.

The book uses theories of industrial policy as the primary base for assessing the ways China conducted its telecommunications revolution. In focused chapters, it also considers theoretical work on central–local relations and foreign direct investment to analyze the successful course of the sector in the 1980s and 1990s, as well as the challenges of the 2000s. In doing so, it builds on these theories by indicating how forces of modernization and the process of globalization affect the policy choices for governments seeking to construct and maintain healthy industrial sectors.

In the case of China's telecommunications industry, the measure of success is: how well has the industry served the population, and satisfied needs for communications tools? What kind of contribution can and does it make to economic advance, to business growth, to educational and social development? Are standards of telecommunications services and equipment production reaching global levels of quality? This study stands in contrast to much of the industrial policy literature (discussed below) that typically gauges success in more narrow terms of international competitiveness. On that level, many parts of the Chinese telecommunications industry that currently lag are making rapid progress, as the book's later chapters will show.

To provide focused analysis of the benefits of a robust telecommunications sector, the book employs parts of the large literature on domestic telecommunications development theory, and examines China in light of changes in other developing nations. Is the country a model for countries that still lag in the communications arena? What do patterns in developed nations indicate for the future of telecommunications growth in the PRC? Or do some factors in the China case make its model difficult to imitate by other nations?

In synthesizing theories from several schools of thought, the book assesses the ways a key industry is shaped in a political system still dominated by a single political party intent on maintaining power. The sometimes contradictory forces of desire for development versus fear of losing control of communications tools also play a role in the way China's telecommunications industry has grown over the past two decades, and later chapters in the text highlight this dilemma.

This chapter outlines the book's analytical framework, and sets the stage for the empirical work that follows. It begins with a discussion of the theoretical literature.

Industrial Policy Frameworks and their Application in Developmental Studies

Industrial policy theory has, for the past several decades, been an important framework of analysis for countries at all levels of development. It is

Industrial Policy

not possible to address all facets of industrial policy theory here, but seminal works do set the stage for analysis of the China case.

One of the most prominent scholars to use industrial policy analysis to account for a nation's rapid economic growth in the twentieth century was Chalmers Johnson. In analyzing the ways Japan successfully built its economy in the years since the 1870s, Johnson built on work by Max Weber and, more recently, Ralf Dahrendorf, Ronald Dore, and George Kelly, to enunciate the idea of 'plan rationality'.[1] In nations that are late to reach economic maturity, the state takes on developmental functions, and, for the sake of rapid growth, eschews dependence on market forces. As its primary tool, the government employs industrial policy, which Johnson called 'a concern with the structure of domestic industry and with promoting the structure that enhances the nation's international competitiveness'. This implied a 'strategic, or goal-oriented approach to the economy'.[2]

In the case of Japan, Johnson's book examined the way many sectors (such as textiles, steel, automobiles, and others) developed over the twentieth century, and profiled key government actors and ministries that shaped their successful growth in both domestic and international contexts. Key elements of industrial policy in this industrial growth context included:[3]

- use of plans to set goals for the economy
- extensive use, narrow targeting, and timely revision of tax incentives
- extensive reliance on public corporations to implement policy in high-risk areas
- creation of governmental financial institutions
- 'the orientation of antitrust policy to developmental and international goals rather than strictly to the maintenance of domestic competition'
- government-sponsored research and development
- 'the use of the government's licensing and approval authority to achieve developmental goals'
- 'administrative guidance', meaning the 'allocation of discretionary and unsupervised authority to the bureaucracy'.

In the international realm, the state, mainly in the guise of the Ministry of International Trade and Industry, or MITI, took measures from the 1960s to delay opening vital and often vulnerable sectors (such as automobiles, steel, chemicals, and electronics) to foreign competition. It also attempted to 'promote large-scale mergers [in these sectors] in order to produce concentrations of economic power' on a par with rivals such as the United

States or West Germany. MITI also sought to limit foreign investment in and ownership of domestic Japanese industries.[4]

Johnson concluded that a model of growth led by the state and its bureaucracy could be applicable to other nations. He suggested 'a different society might be able to manipulate its own social arrangements in ways comparable to those of postwar Japan in order to give top priority to economic development...'.[5]

Johnson did see shortcomings in his scheme. 'The fundamental problem of the state-guided high-growth system is that of the relationship between the state bureaucracy and privately-owned businesses.'[6] Undesirable outcomes included, in various instances, domination of industries by state-sponsored cartels and inhibition of competition when corporate management comes under state control.[7] However, he found Japanese methods 'preferable to either pure laissez faire or state socialism as long as forced development remains the top priority of the state'.[8]

Scholars of East and Southeast Asian development were quick to pick up on Johnson's arguments, and to analyze other nations in their light. Stephan Haggard and Tun-jen Cheng, for example, considered South Korea, Taiwan, Hong Kong, and Singapore—also known as the 'Four Little Tigers' (or sometimes 'Dragons')—in the context of a state-guided development model. They found that (except in Hong Kong) the state played a major role in developing pragmatic trade policies toward foreign investment as well as export-oriented growth.[9] The governments 'played a central role, supporting the activities of business, targeting particular sectors for investment, and engaging in production of heavy industries such as steel'.[10] Ezra Vogel also analyzed the 'four dragons', and asserted that, partly for historical reasons, Japan's model was of great importance.[11] Among other factors that Vogel found shaped development in the region was a meritocracy of rule that meant state bureaucrats 'played a critical role in industrialization'.[12]

In 1993, the World Bank, until then an advocate of reliance on markets in the mode of neoclassical economics, and suspicious of state intervention in economies, published a report recognizing the important role of government in fostering industrial growth.[13] The study, which focused on Asian nations in East and Southeast Asia, indicated the state could and should play a role in areas such as absorbing foreign technology, regulating financial systems, and building human capital.[14] However, the World Bank found a mixed record for nations trying to promote specific sectors using industrial policies. Only Singapore and, to a lesser extent, Japan, showed positive growth, measured in the somewhat narrow terms of 'total factor productivity' (TFP) in industrial sectors targeted by the various governments.[15] Still, the World Bank, while not repudiating the neoclassical model, did suggest industrial policy helped produce positive results for sectors in export-oriented areas.[16]

Sanjaya Lall faulted the World Bank study on several measures, in particular for using the rather imprecise measure of TFP to gauge the success of industrial policy.[17] Lall suggested that useful ways to judge industrial policy effects included 'the rate of growth of exports by the country, the skill or technological composition of exports and production, indices of local technological development...[and] research and development'.[18]

Lall's own work focused on the ties between industrial policy and technological development and foreign direct investment in developing nations. He asserted that governments needed to mount interventions to overcome market failures, and that technology policy needed a 'strategic vision'. Furthermore, countries should develop outward-oriented trade regimes to force firms to compete in world markets. Technology infrastructure also plays a key crucial role in supporting industrial technology. However, countries should avoid creating 'hothouse' technologies (ones that grow under excessive protectionism) with little industrial use.[19] The Asian 'tiger' economies also prospered as governments were able effectively to channel foreign investment, though in their own unique fashions, to strengthen areas touched by foreign capital and technology.[20] Lall's work on Taiwan, South Korea, Hong Kong, and Malaysia all pointed to the effectiveness that government actions could have in fostering key industries and encouraging international competitiveness.

Scholars examined other areas of the world, and found more difficulty in successfully applying the industrial policy model. In Latin America, for example, Peter Evans pointed out many structural differences between nations in the two regions. The United States long sought to 'protect the economic status quo' in the neighboring nations to the south, with the result that extractive industries, agricultural or mineral, dominated many of the region's economies.[21] American multinational corporations thereby contributed, in both the pre- and post-World War II eras, to a climate of dependency in which national governments played only a passive role in choosing which industries would be of major benefit to the economy. Evans suggested that 'a long, unbroken historical experience of foreign direct investment produces a greater likelihood of inequality', thus making the task of maintaining popular support for government economic policy more difficult.[22] A lack of land reform in Latin America and domination of powerful rural families further inhibited the ability of the state to formulate coherent economic growth strategy.[23] This contrasts to the cases of East Asia, in which each major nation has either seen conscious land reform or, in the cases of Singapore and Hong Kong, historically lacked a significant rural sector.

Evans concluded that nations such as Brazil, Guatemala, or El Salvador, because of these historical and resulting structural differences, would have difficulty emulating the state-led industrial growth paths of nations such as Taiwan or South Korea. He did suggest, however, that countries such as Cuba or Nicaragua, which have had 'highly autonomous state apparatuses and thorough land reforms', might be in a better position to emulate the Taiwan example.[24]

Other more recent studies have borne out some of the difficulties of transplanting the East Asian industrial policy model outside the region. Marcela Miozzo found, for example, that transnational corporations confounded Argentine government attempts in the 1990s to create a favorable environment for growth of its automobile industry. Foreign corporations, allowed to play a large role in the industry through regional trade agreements, passively inhibited modernization of both production restructuring and technological upgrading, and left Argentina with a less competitive vehicle sector.[25] The implication is that a firmer state role in the development process would have led to a more efficient vehicle industry. Lall and Ganeshan Wignaraja found the situation even grimmer in sub-Saharan African nations. Across the region, and even in such relatively enlightened nations as Ghana, low levels of technological skills among the population greatly complicated attempts of government policy to spur economic growth.[26] In India, restrictions on foreign direct investment, limitations on growth of large firms, restrictions on licensing of industry, and other counterproductive measures until the 1990s retarded economic growth and technological development.[27]

Pitfalls of Industrial Policy

Following the beginning of Japan's recession in the early 1990s, and in the years leading up to the Asian economic crisis of 1997, some scholars began to point out problems of the development model even for nations where it had seemingly had success. Paul Krugman echoed the earlier-cited World Bank study by reasserting the shortcomings in total factor productivity in the 'Asian tigers' as well as China, though Seiichi Masuyama and others questioned his way of calculating economic productivity growth.[28]

Industrial Policy

Masuyama suggested that future growth would have to be 'derived mostly from technological innovation, rather than from the traditional source of labor and capital inputs'.[29]

The 1997 Asian financial crisis led to some further criticism of the utility of industrial policy. Popular press accounts such as those in *Business Week* asserted 'the Asian crisis proves industrial policy doesn't pay'.[30] Complaints included the assertions that industrial policy prevented the market from weeding out failures, that protected banks kept bad loans on their books, and that corruption resulted from crony capitalism.[31]

François Godement called the crisis 'Asia's major economic emergency of the past 50 years'.[32] However, he asserted that Asian productivity (including the TFP measure) had been rising in the 1970s and 1980s, in particular for technology sectors.[33] For him, the main cause of the crash was inefficient government monetary policy, particularly in Japan, along with regional difficulties in economies adjusting to trade liberalization.[34] His policy recommendations, even at the height of the crisis, hesitated to criticize the state's role in building economies (except for prevalence of crony capitalism in some nations), and instead prescribed measures for debt relief and banking reform.[35]

Even without many of the systemic financial changes that Godement and others suggested in the late 1990s, within a few years many of the nations most affected by the 1997 events had seen some kind of economic rebound. By the fourth quarter of 2002, for example, South Korea reported gross domestic product (GDP) growth of 6.8 percent over the previous year, Taiwan was up 4.2 percent, and Malaysia grew 5.6 percent. Industrial production was also ahead in most of the tiger economies, with early 2003 seeing growth averaging about 5 to 6 percent in South Korea, Singapore, and Malaysia.[36] Though these are rough measures of success, they do indicate that the worst of the late 1990s problems were not crippling in the longer term.

China was an exception to the 1997 crisis. Unlike the nations in Southeast Asia, it did not devalue its currency, nor did it suffer GDP loss. In fact, its annual GDP growth rate kept close to 10 percent in nearly all the years from 1997 to 2002.[37] Unlike most of the Asian tigers, China was able to avoid significant currency devaluation, and maintained a strong export record even in the face of challenges from its weak-currency neighbors.

Before turning to the case of China, and focusing on the telecommunications industry, we can come to some general conclusions on industrial policy for the purposes of this text. On the positive side, industrial policy in the East and Southeast Asian context, despite some shortcomings, was effective for fostering growth of key industries in several of the region's nations. As the above-cited studies indicate, Japan, Taiwan, South Korea, Hong Kong, Singapore, and, to a lesser extent, Malaysia, used industrial policy in ways to target vital parts of the economy for funding and nurturing. These policies allowed the selected areas to become important parts of the domestic economy and even competitive in global markets.

However, we must also remember the problems that can be associated with inappropriate government intervention. As noted above, Johnson warned that internal inhibition of competition could foster cartelization and industrial inefficiency. Evans, in a later study, warned of predatory states that 'extract at the expense of society, undercutting development'.[38] However, Evans juxtaposed the 'predatory state' against the 'developmental state', asserting that the negative affects of state intervention are alleviated when a set of social ties binds the state to society, and allows a flexible negotiation of goals and policies (in what Evans labeled 'embedded autonomy').[39] Evans's comparative study focusing on information technology industries found Korea to have a version of this ideal type of state intervention, while Brazil and India were only partially successful.

Overall, it is difficult, if not impossible, to prove that any sort of government intervention brings about better results than little or no state involvement. As Howard Pack and Kamal Saggi argued in a 2006 essay published by the World Bank: 'We do not know how Japan would have fared under laissez-faire policies... [i]t might have done still better in the absence of industrial policy—or much worse.'[40] They asserted, however, that the process of governments picking key sectors is 'exceptionally complex', and that, particularly for developing nations open to foreign capital and investment, it may be more efficient to 'allow foreign firms to facilitate cost reduction in the host economy'.[41]

Andrew Cortell's work summarized some of the key issues for evaluating industrial policy effectiveness by suggesting three straightforward criteria. First, were the policy's objectives achieved? Second, did the policy's costs outweigh the benefits for the industry? And third, what was the nature of the policy's spillover into related industries, and for the country's relations with other nations?[42] In his own cases of British and American efforts to foster information technology component production, Cortell found greatest success when industrial representatives worked with the government to provide information on which technological goals were most achievable, and when programs sought to meet but not necessarily surpass capabilities of competitor nations' industries. Shortcomings appeared when policies dictated overly ambitious results.[43]

Industrial Policy

This study, then, keeps in mind the road maps provided by Johnson, Evans, and others for pinpointing the contributory measures of industrial policy, while watching for the negative effects that state intervention may have on a particular industrial sector. The analytical chapters of the book hold up the salient points of these models to the telecommunications industry in China.

As the chapters will show, up to the first decade of the new millennium, the telecommunications industry appears to have been a broadly successful beneficiary of industrial policy in the People's Republic. In post-WTO China, however, the policies may need to be adjusted to account for a new global economic environment.

This is of course not the first study of industrial policy impact on China, and, as it turns out, industrial policy has not had uniformly positive effects. Before turning to the substantive analysis of the industry, I consider other studies and findings of work in different sectors of the Chinese economy.

Industrial Policy in China

Dorothy Solinger conducted one of the best early studies on general industrial policy in China. She examined the first years of the country's reform era, from 1979 to 1982, a time when new leaders were faced with challenges left by the disruptions of the previous radical regime of Mao Zedong. She found a prerequisite foundation of state control over the population, ability of the government to mobilize capital and credit to achieve its goals, and a bureaucracy 'primed to effect developmental goals'.[44] In comparing China's social context with other nations, she noted 'Confucianism's emphasis on obedience and on dependence on authority has influenced Japan as well as China'.[45] She concluded that 'industrial policy, in short, has structural regularities that transcend ownership systems, mode of production, institutions, and political forms, if certain societal traits are present and if a specific conjunction of historical forces and economic capabilities exists'.[46]

In addition to the traits highlighted by Solinger, China also has several conditions for successful industrial policy implementation noted in the previous section. Unlike Latin America, for example, China faces few problems of challenges from rural, landed elites, as land reform in the early 1950s effectively muted this potential challenge. Furthermore, following nationalization efforts in the early and mid-1950s, the country has avoided the trap of foreign multinational corporations dominating and shaping the economy in ways that might be detrimental to domestic industry. China, then, seems to fit quite comfortably into the mold of a strong state with a guiding bureaucracy capable of nurturing key sectors, in the same league as Japan, Taiwan, South Korea, and Singapore.

In supporting her assertions that industrial policy could be effective in China, Solinger's methodology centered around a case focus on the city of Wuhan. Here, she focused on city as well as central policy planners, and their ability to shape municipal sectors ranging from textile manufacture and other light industry to machinery, metallurgy, and other heavy industry.[47]

Several other China scholars have examined the course of industrial policy using empirical evidence from one or more specific industrial sectors. Kenneth Lieberthal and Michel Oksenberg considered the way policy was made in the energy industry, and used the sector to highlight issues of top government leadership, central–local relations, and the role of foreign investors.[48] More recently, in considering problems of state-owned enterprise reform, Edward Steinfeld presented a compelling discussion of the steel industry, and pointed out important issues of clarifying property rights and instilling 'hard' budgetary regimes to enhance productivity in an old-line industry.[49] Thomas Moore thoroughly examined the textile and shipping industries, and found that forces of global surplus capacity and the international Multifiber Arrangement regime, rather than direct state intervention, were important factors in industrial change.[50] Adam Segal's groundbreaking study of non-governmental high-technology enterprises highlighted the contribution that local governments could make in nurturing smaller corporations.[51] And Daniel Lynch's text on Chinese media industries gave a brief though useful overview of the telecommunications industry in the late 1990s, pointing out the dilemma of improving communications services versus maintenance of political control.[52]

My own study of the automobile industry in China did not specifically judge the utility of state intervention through industrial policy. However, those companies that received the most government support, in particular at the municipal level, such as the joint venture Shanghai Volkswagen, were the most successful in expanding production and modernizing their facilities. Those with less central and local government support, including Guangzhou Peugeot and Chrysler's Beijing Jeep Corporation, became minor players in China's passenger car industry. Finally, I found that those Chinese state-owned companies best able to cooperate with a foreign investing partner ended up as the most productive.[53]

In many instances, these and other works have pointed out both the successes and failures of Chinese government policy. Steinfeld's work, for example, criticized a lax state financial system, one with banks that 'neither effectively select borrowers nor effectively monitor funds once dispersed'.[54] In this environment, the state could play an important role, not by intervening, but by reforming the system to make firms more responsible under 'hard' budgetary constraints.[55] However, he asserted that 'where basic market-like constraints already exist, state intervention—particularly in the area of resource allocation—becomes either superfluous or patently destructive'.[56]

Moore took more direct issue with the developmental state model, and suggested that 'most Chinese "success stories" during the post-Mao era [as with his own two case studies] have involved industries where state intervention has declined'.[57] However, Segal argued that 'government support remains crucial to growth in China', but, in his focused arena of small enterprises, found local governments and decentralized state-affiliated institutions more important for shaping development than central policy.[58]

One problem with case study analysis is that it must be careful about extrapolating results beyond the case or cases under study. Thus while Moore, Steinfeld, and others have perfectly valid conclusions about the areas they examined in detail, their assertions may not necessarily be valid for the larger universe of Chinese enterprise.

Other scholars have chosen the telecommunications industry for assessment, but have generally used either descriptive or market-oriented analytical frameworks, rather than industrial policy approaches. The following section considers some of these studies.

Studies of China's Telecommunications Sector

Most of the major early works examining telecommunications development in the PRC came from scholars in the communications or economics fields. One of the first and most comprehensive studies of the sector was a text by communications specialists Milton Mueller and Zixiang (Alex) Tan.[59] Their 1997 book *China in the Information Age* considered central–local telecommunications relations of the era, and noted problems of tight government control within the telecommunications ministry, as well as over the growing Internet and foreign investment in equipment and operations. This important book gave a wonderfully concise snapshot view of key policies and developments from the late 1980s to the mid-1990s.

Major studies from the early 2000s continued assessment of policy development to the end of the 1990s, and mainly focused on the deployment of fixed-line and mobile phone networks. For example, economists Ding Lu and Chee Kong Wong used a market-oriented approach to study investment and expansion decisions in constructing China's telephone network. They considered major rules and regulations developed to shape the sector, and assessed supply and demand factors that led to the rapid growth in the number of telephones accessible to the increasingly wealthy population.[60] Xu Yan and Douglas Pitt, both based in the communications field, gave a thorough and highly detailed analysis of a decade of regulations, ones also mainly related to the spread of fixed and mobile telephones. Their study emphasized the virtues of deregulation in the telephone sector, and indicated that trends toward industry privatization and global competition presented both market problems and opportunities for the voice communications companies. [61]

As the Internet became a major phenomenon in both the developed world and in China in the 1990s, scholars produced works examining the growth and impact of the new data communications technology. As with studies of general telecommunications, early writers were based in the communications discipline. For example, Zixiang Tan and Wei Wu wrote on the early growth and expansion of the data network.[62] Their work gave a thorough technical discussion of the Internet's first years, and served as a useful guide to some of the intra-governmental rivalry as the network grew in the early 1990s. However, because of their base in the communications arena, these writers generally avoided discussion of other sociopolitical factors such as censorship, use of the network for information distribution, and regional access discrepancies. Other earlier writers, such as Bryce McIntyre, focused almost exclusively on the network's hardware.[63]

With the spread of the Internet in China in the 1990s, communications specialists were joined by a new variety of social scientists seeking to understand how the new data tools could affect broader development of China's political and social landscape. One of the first studies was by Geoffry Taubman, who tried to put Internet development in a context of social change and threats to government control. He asserted the Communist government's hold over domestic affairs would eventually be diminished because of the new technology.[64] Political scientist Daniel Lynch also considered the ways the government controlled and censored the Internet to the late 1990s, and suggested that viewers could in fact

Industrial Policy

bypass government censorship of foreign websites via a variety of methods.[65]

Studies of China's Internet in the early 2000s continued a focus on issues of social and political content regulation, and often highlighted censorship policies. In 2002, Michael Chase and James Mulvenon provided a first-rate analysis of ways dissidents could and did use the Internet to express controversial viewpoints, and chronicled ways the government reacted to their challenges.[66] Shanthi Kalathil and Taylor Boas argued that, while the government had the upper hand for the moment, its drive for economic openness would create future opportunities for political expression on the network.[67]

As for the analytical works on either telecommunications or the Internet published in Chinese for a domestic PRC audience, nearly all have tended to be relatively policy-neutral and politically circumspect—the political environment under which most China-based Chinese scholars function has shaped many such writings in this way. Perhaps the best text on telecommunications was Wang Xiangdong's 1998 book *Informatization: China's Choices in the Twenty-first Century*, which included focused attention on basic government policy as well as the information network's technical growth. [68] Wang's book, however, omitted most discussion of the political conflicts related to the network's development, and had little on the social implications of the data communications web's growth.

In focusing on the broader advance of telecommunications network construction and operation, as well as the advance of the Internet, over the past several decades into the beginning of the twenty-first century, my own study fills several gaps in the political, economic, and sociological literature. My work synthesizes the ways the government, at both the national and local levels, built, managed, and has come to control one of the largest communications networks in the world.

In the following section, then, I outline my own approach and arguments about the development of the telecommunications sector in China. In taking account the role of the state, I examine whether criticism of industrial policy in China can or should be made as a universal argument.

The Argument

As implied in the above discussion, one of my premises for assessing the utility of industrial policy is that the results of state intervention can be measured on an *industry-specific* basis. That is, an interventionist state may use similar methods, but the end result could be positive for some sectors, and negative for others.

This approach naturally can attract the charge of being reductionist, and in fact one does lose some analytical power in the construction of the framework by drawing narrow conclusions. However, this kind of industry-specific analysis allows us to avoid the pitfalls of work that, finding industrial policy failure in one or several sectors, indicates that it has little or no utility, and should be avoided.

This argument builds on the comprehensive work of Richard Katz, who described Japan's current system as a 'dual' economy. In his analysis, the strongest sectors of Japan's economy were automobiles, consumer electronics, semiconductors, and machinery, and other areas that had been infant industries and 'needed an initial jump-start' supplied by government promotion and protection.[69] The weakest were food processing, textiles, petroleum refining, and other sectors that were shielded from international competition.[70] By the 1970s and 1980s, as all parts of the Japanese economy had matured, the continued protection fostered inefficiency across a broad spectrum of the economy. In essence, industrial policy was useful and effective for new industries that needed government support to establish themselves in Japan and in the global economy, though it did become a negative influence once the industries had matured.

In a similar way, I argue that industry-specific study of China should be careful not to draw broad conclusions based on one or two industrial sectors. For example, Steinfeld was exactly right to criticize state intervention in the steel industry, as he showed with great accuracy how the system of soft budget constraints led to a suboptimal outcome. However, it may not be correct to reject the utility of industrial policy for fostering sectors that can effectively use a 'jump-start' to become valuable parts of the country's economy.

This book will show that industrial policy in China's telecommunications industry, at least to date, has been a case of successful industrial policy. It will analyze several different components of the sector, and indicate that, after more than twenty years of effort, the industry has progressed faster than nearly any other such industrial sector in the world.

The following chapters present several ways of assessing the positive and negative feature of China's industrial policy achievements in the telecommunications sector. My scope of analysis includes telecommunications services industries, such as basic fixed-line telephones, mobile phones, along with the Internet. A starting point for measuring whether this part

Industrial Policy

of the sector has been successfully grown is examining the spread and accessibility of telecommunications. How many people have access to a telephone or the Internet? Has the quality of service improved under government guidance? Has the growth been sustainable, with the industry avoiding irrational costs to grow the sector? In other words, did the Chinese government create a functional, viable, and high-quality industry that is on a path to be competitive even in a post-WTO global competitive environment?

Chapter 2 begins by putting the Chinese case in the context of telecommunications development trajectories. It briefly considers policies in other countries where growth was in many cases slower, and where government intervention (except in the cases of Japan and South Korea) was less effective. The chapter next examines historical developments in the Chinese sector up to the beginning of the reform era, ones that set the stage for effective industrial policy beginning the early 1980s.

Chapter 2 then turns to the post-1976 reform period in China, and chronicles the ways the government developed policies to foster the industry. It discusses policies employed that were strikingly similar to what Johnson described in the Japan case of the 1950s and 1960s: favorable grants and loans, effective tax policies, shielding of the sector from international competition, and other strategies that encouraged the industry to expand rapidly.

A key part of the analysis of the sector comes in Chapter 3, which asks the question: can and do companies owned by the state actually effectively compete with each other? As Johnson and others have pointed out, a major potential problem with the shielding effect of industrial policy is that state-owned or -protected corporations have no reason to strive for greater efficiency, better service, or a larger market share when they are essentially all part of the same family. However, the very important examples of the giant public Chinese telecommunications service companies China Telecom, China Unicom, and (later) others, which all became fiercely competitive, show that industrial policy does not necessarily breed slackers. This chapter examines reasons for this competitive nature, and assesses the positive benefits for the Chinese economy and for consumers.

Chapter 4 considers several dimensions of China's Internet. It begins by examining various other national models for Internet development, and puts the PRC in a comparative context. It then assesses in turn the state role in building and managing the network, introducing Internet service companies, and overseeing the growth of mainly private content providers. While the chapter considers issues of censorship and control in a way similar to earlier studies noted above, it also highlights the dilemma the government faces of wanting to use the network for economic benefit, while also wanting to keep control over political challenges. As such, it details major developments in e-commerce from the 1990s to the mid-2000s. Finally, the chapter also notes the evolving demographics of network users, and assesses how the 'netizen' profiles shaped usage patterns.

My definition of the telecommunications sector goes beyond the service industry, and includes equipment that can be used to broaden the ability of China, as well as citizens in potential export markets, to use instruments of communication. Chapter 5, then, focuses on the country's telecommunications equipment manufacturing industries, and assesses the way industrial policy shaped it. It examines several key companies to see how well they have satisfied domestic needs, whether the companies have achieved technological advances, and how competitive they are in the international marketplace. Here we can use some of the more traditional tools of assessing industries, taking into account sales profits and losses, and the impact of China's WTO entry on industries that have been both shielded from international competition, and benefited from foreign direct investment.

Chapter 6 extends the question of industrial policy to the local government level. In building on earlier work by scholars such as Segal and Eric Thun, it focuses on the city of Shanghai to see how regional governments have acted to build various aspects of the industry.[71] It probes for ways officials outside of the central system have contributed to building the sector on a regional basis.

We should not be satisfied with basic numbers, however, unless we consider how they are distributed across society. Has the industrial policy contributed to larger national prosperity, but been applied unevenly? Are some members of the nation, either regionally or on a class basis, left without the benefits? Chapter 7 examines China's 'digital divide', and looks for a shortcoming in a policy that may have left some of the nation's citizens without deserved access to the fruits of the industrial policy.

Finally, Chapter 8 proposes corollaries to basic industrial policy theory, ones based on the outcome of China's telecommunications transformation. It indicates points other developing nations should heed should they target key sectors for focused state economic policy intervention.

2

China's Telecommunications History and Policy Trajectory

As nations build their telecommunications network, government involvement varies in a wide range of forms. As this chapter will show, China's government was a primary player even before the Communist movement won its revolution in 1949.

Before turning to an overview of the telecommunications history in China, this chapter assesses the ways other nations have built their telecommunications regulatory models. It considers the mix of public and private ownership in operating companies, for both developed and developing nations. What paths existed that could guide the way China could (and still can) shape its developmental course?

The chapter then examines the Chinese case in light of these methods of building communications networks. It considers the sector's growth in imperial times, as well as in the years leading up to the Communist revolution. The modern era is marked by revolutionary leader Mao Zedong's death in 1976, and the rise of Deng Xiaoping and a new generation of technocrats in the following decades. The chapter concludes by examining the role government policy played in setting the course for rapid communications growth at the end of the twentieth century.

Patterns of Telecommunications Growth

Developed and developing countries have followed similar patterns as their telecommunications networks have grown. The capital-intensive nature of building a network has meant that a national government will typically allow monopolistic operation of the network for many years or decades. The monopoly company is either a private company that is regulated by the government as a utility, or a corporation owned directly by the government.

The United States was one of the first nations to create a telecommunications network. Its first effort, construction of a telegraph network, was initially marked by chaos, as some thirteen different private companies operated in five northern states in the early 1850s. Costs were high when customers tried to send messages from one system to another, and the companies were slow to repair damaged lines. The companies could not unite behind uniform rules and practices. By the end of the decade, however, the Western Union Telegraph company had linked many of the disparate lines, bringing rapid expansion and reliable service. By the 1880s, Western Union had acquired near monopolistic control over American communications.[1]

The US government considered creating a rival public company to Western Union, but the company argued that under the national Constitution, the government should regulate, but not take part in, the commerce of telegraph business. Proposals to set up a rival public company failed in the 1860s and 1880s.[2]

Western Union made an early foray into the telephone business in 1877, but was soon in competition with the Bell Telephone Company.[3] In 1879, Bell agreed to stay out of the telegraph business, and Western Union gave up its telephone operations. Bell also brought equipment manufacturing into the company in 1881, when it bought a controlling interest in what became the Western Electric Manufacturing Company from Western Union. In 1885, the company was incorporated as the American Telephone and Telegraph Company, or AT&T.[4]

In 1893, AT&T's patents for telephone technology expired, opening the doors to rival carriers. In competition against independent phone companies, AT&T was sparked to expand its service, and subscriber numbers grew from 260,000 in 1893 to more than 3 million in 1907.[5] Though AT&T expanded quickly, the mainly small, regional telephone rivals continued their challenge. AT&T began to dispose of competitors by simply buying and adding them to the Bell System network. In 1909, the company even bought a controlling stake in Western Union.

AT&T used another tactic against potential rivals: it blocked them from connecting to the national network, thereby isolating them in their own communities. Between 1904 and 1919, thirty-four states reacted to this tactic by passing laws to force interconnection.[6] In 1910, the federal Interstate Commerce Commission (ICC), regulator of the telephone business, investigated whether AT&T was violating antitrust laws, and becoming a monopolistic service. The so-called 'Kingsbury Commitment' of 1913 ended the company's short-lived controlling stake in Western Union. It also indicated AT&T would no longer buy small rival companies, but would allow them to interconnect to the national system.[7]

AT&T also controlled the bulk of equipment manufacture through its affiliate Western Electric, which sold devices to competitors only if necessary. Equipment manufacture advanced, however, even under these near monopolistic conditions, as the phone company merged its research and development staff with that of Western Electric in 1907. The company formally established the Bell Telephone Laboratories in 1925. These soon would evolve into one of the world's greatest innovating bodies.[8]

For nearly the rest of the century, the government let AT&T maintain near monopolistic control over the vast network it had built and which continued to grow. The US government did regulate communications rates and services through the ICC and, after passage of the Telecommunications Act of 1934, the Federal Communications Commission (FCC). Not until 1984 was AT&T forced to split into several competitive local phone companies. But the FCC continued as the government regulator for many aspects of telephone and other communications services.

Despite AT&T's dominance, a significant role for the federal government did emerge in the middle of the twentieth century. AT&T was reluctant to provide service to America's rural areas, as more lucrative markets could be found in densely populated urban communities. By the late 1940s, only 38 percent of US farms had any kind of telephone service, and what did exist was of antiquated form.[9] In 1949, the federal Rural Electrification Administration (REA) was therefore empowered to provide loans to develop rural communications, and funds flowed to both rural cooperative ventures and independent telephone companies.[10]

By 1964, the REA had made loans totaling some $1.1 billion to more than 800 rural companies and cooperatives, with the result that 79 percent of farm homes had at least some type of telephone service. However, about 65 percent of the loan funds were targeted for the private telephone companies, and only 35 percent went to the cooperative ventures.[11] A 1973 study indicated non-Bell System companies provided service to 59 percent of the nation's land area (excluding Hawaii and Alaska), though AT&T, with its concentration in cities, owned 83 percent of American telephones.[12]

Though the government had taken a major role in regulating the growth of the communications network, and had even directly subsidized development in rural areas, the American telecommunications network in the twentieth century essentially developed as a private system with one major corporate player. Most other currently developed nations followed a similar path of monopolistic control, but with a public, typically national government, owner.

Great Britain and France, for example, developed telegraph and later telephone services using public monopolies in the late nineteenth century, and both used their Post Office departments to run their networks. Britain nationalized private telegraph companies in 1868, and in 1911 took over the bulk of private telephone companies.[13] The phone network grew slowly under the national monopoly, as the government was reluctant to add capital investment to expand and improve the network. Only in 1981 was British Telecom (BT) established as a state-owned company independent from the Post Office, and privatization took place over the coming decade and into the early 1990s.[14]

France likewise outlawed private telegraph companies in 1837, and nationalized the nascent private telephone system in 1889. As with Britain, the government was reluctant to finance the network.[15] By 1900, the nation had only 30,000 phones; in contrast, in 1909 the 100 largest hotels in New York City had 27,000 telephones.[16] Up to the mid-1970s, the French government failed to provide needed investment in the sector, leaving the nation with a poorly functioning network. Only in the late 1970s and 1980s did the government begin to increase funding for telecommunications, and the state-owned operating company was renamed France Telecom (FT) in 1988.[17] Like BT, FT was gradually privatized over the next fifteen years. Both the British and French companies also came to see domestic competitors; the issue of competition is discussed in Chapter 3.

Like their European counterparts, developing nations were generally slower to relinquish government monopolistic control over telecommunications service and equipment manufacture. Foreign companies, usually with superior management skills and technology, were kept out of not only operating telecommunications systems, but of investing in equipment manufacture. For many developing countries, official corruption further served to hinder growth of important public utilities such as telecommunications.[18]

Some Latin American countries gradually moved toward liberalized regimes. Mexico's TELMEX phone company initially grew under foreign private ownership beginning in the 1920s. But Mexican private shareholders took over the corporation in 1958, and the Mexican government bought control of the service in 1972. Under inefficient central government management, the spread of phone lines slowed—as of 1989, only 16 percent of households were connected to the public network. The following year, the government re-privatized the monopoly. The terms of privatization included stipulations for rapidly upgrading services, and introduced a rival carrier, TELECOMM, to provide competition.[19] The government even allowed foreign companies a share, with France Telecom and the American company Southwestern Bell taking minority stakes.

Following the privatization, Mexican billionaire Carlos Slim moved to expand the network, and by early 2005 Mexico had about 52 phone lines per 100 people. However, much of this growth took place in the mobile phone sector dominated by América Móvil (also controlled by Slim). TELMEX continued to control 95 percent of the country's fixed-line network, which grew more slowly.[20]

Argentina, in contrast, failed in several attempts in the 1980s to privatize its own inefficient state carrier.[21] As of 1990, the waiting list for telephone connections could be many years.[22] But that year, two consortia of European phone companies created a duopoly for phone service in Argentina.[23] By 2001, the number of lines had nearly tripled, to 8.6 million for 37 million people. The number of cell phones grew from 15,000 to almost 7 million over the same decade.[24]

Parts of Africa saw similar trends.[25] In 1977, Kenya created its Posts and Telecommunications Corporation (KPTC). The government agency had a monopoly over all telecommunications services, as well as radio frequency allocation and postal matters. In the early 1990s, reforms allowed private companies to make equipment—but the monopolistic carrier continued to purchase almost all of the manufactured products. In Nigeria, too, the government owned and managed a limited liability company. In the late 1980s, moves to raise rates up to 7 times their previous levels led to great public dissatisfaction. Kenya, however, dismantled and privatized its monopoly telecommunications corporation in 2000, and by 2004 had two thriving rival mobile phone corporations.[26]

In Asia, India—like China a large nation with a significant rural population—had a small and inefficient communications system at its independence in 1947. Telephones existed only in the cities, and there were few programs for extending them to rural areas.[27] The British left their colony with only 82,000 working lines for some 350 million people.[28]

The independent government's goals, stated in five-year plans over the following decades, saw communication as a tool for improving education, encouraging civic cooperation, and raising living standards.[29] But a reluctance to privatize the industry, failure to import advanced equipment, and low investment levels retarded the sector's growth. In the late 1980s, new policies indicated government agencies would cooperate to utilize more advanced technologies, and focus attention on rural areas. The monopolistic Posts and Telegraph Board was partially split in 1986. But a new Department of Telecommunications continued to play the roles of regulator, service operator, and equipment manufacturer, and protests by unionized government workers prevented meaningful competition or foreign participation in the industry.[30]

In 1993, there were still only about 8 million phone lines for India's 900 million people. But the country finally began allowing domestic private-sector competition for services and foreign investment in manufacturing in the 1990s.[31] The result was a rapid expansion of the nation's network, and more than sixfold subscriber rate increases from 1995 to 2003.[32]

Two East Asian nations that, as we have seen, employed industrial policies in various sectors in the twentieth century also saw a major role for the state in the development of their telecommunications industries. We consider briefly the ways Japan and South Korea developed their communications networks over past decades.

As discussed in Chapter 1, the state has played a major role in many aspects of Japan's modern economic development. Many of the roots of industrial policy are found in nineteenth-century Japan, following the rise of reformist leaders of the Meiji imperial restoration in the 1870s.[33] Japan was one of the first Asian nations to develop a communications network, and the government took the lead to introduce the first telegraph line from Tokyo to Yokohama in 1869, and to connect all major cities with lines by 1880.[34] The country's first telephone service opened in 1877, and was supported by a short-term foreign loan taken by the government.[35] In 1885, the government created the Ministry of Communications to oversee the spread of telephone and telegraph networks.[36]

The state formally took ownership of the network in 1889, though there were periodic calls for privatization. The growing role of the military in Japan's political system, however, kept this vital sector under government control. The Ministry of Communications, which both operated and regulated the network, made plans to create a half-government, half-private company in the 1930s, but they were stopped by an economic downturn and loss of subscribers in the 1930s.[37] Equipment manufacture, however, came to fall into private corporate hands: a joint venture with America's Western Electric contributed to the growth of the Nippon Electric Company (NEC). By 1911, most parts were made by private companies in Japan.[38]

Overall, the state-run network grew at a relatively good pace in its first few decades. By the mid-1930s, the density of telephone extensions in Japan was 1.9 per 100 people, still significantly less than that of the US (at about 15.1), but not so far behind Western Europe (at 7.1).[39] In fact, by this time the spread of telephones in Japan was roughly the level of both India and (as we will see later) China in the early 1990s.

Public ownership of the network continued after World War II. In 1952, the Nippon Telegraph and Telephone Corporation (NTT) was established as a government-owned monopoly for domestic communications services. NTT would continue to dominate most services for the next several decades, until it was broken up in the 1990s—this will be discussed briefly in the next chapter. The new Ministry of Posts and Telecommunications served as the industry's regulator, and the Japanese parliament and the Ministry of Finance regulated NTT's budget and investment policies. Foreign countries were prohibited from taking ownership in the company.

The government did provide for some non-state role in the industry, as it gave a monopoly over international communications to the private company KDD (Kokusai Denshin Denwa). Still, the Ministry of Posts and Telecommunications continued to regulate both companies.[40] The sector grew with Japan's overall economic success, and by the late 1960s telephone penetration in Japan, at about 17 per 100 people, had nearly reached the levels of Western Europe.[41]

South Korea's telecommunications sector had a similar strong role of the state, as well as monopoly control in the first several decades. Korea's colonial development under Japan (from 1910 to 1945) set the stage for its political organization, and the independent country's Ministry of Communications both operated and regulated the industry until the early 1980s. In 1982, however, it split off the Korea Telecommunications Authority, a state-owned corporate entity renamed Korea Telecom in 1990. As in Japan, the Ministry continued as the sector's regulator.[42]

From the 1980s, the Ministry of Communications played a vital role in expanding the nation's network. The Ministry not only increased public funding for communications services, but also established several new agencies to promote research and development, telecommunications policy research, and ways of using information in the larger society.[43] Privatization of Korea Telecom began in the early 1990s, and competition for long-distance service emerged at the same time.[44] South Korea's telephone penetration rate had almost matched that of the former colonizer by 1991, when the country had 34 phone lines per 100 people, compared to Japan's 45.[45]

In these brief analyses of Japan and South Korea, we see some evidence that a strong state role can benefit the populace through a relatively rapid spread of telecommunications links. Government emphasis on building communications networks was key to the expansion of services, and resulted in the two Asian nations rather quickly coming close to matching penetration rates of developed nations.

In general, both developing and most developed countries have historically tended to follow similar patterns.[46] First, they put overall control of the sector in the hands of a central government agency, one that is sometimes combined with postal services to create the generically named 'Posts and Telecommunications Administration' (PTA). The PTA is usually at a ministry or executive agency level in the political hierarchy, is the monopoly operator, and sometimes is also the sole regulator and equipment manufacturer. Through the PTA, the government can guarantee uniform standards, coordinate purchases of equipment, and regulate development in all areas of the nation. The existence of a PTA, though, means competition is suppressed. Moreover, it is public bureaucrats, rather than private entrepreneurs, that usually manage the industry.

A second feature of telecommunications development is related to the first: the monopolistic agency tends to keep out foreign competition. Often, issues of national security are raised to defend this position—but the lucrative income of operating a communications system (and sometimes manufacturing its equipment as well) is often a tempting target of protectionism.

Finally, developing nations often focus on urban growth in their early stages of telecommunications deployment. Large cities are natural targets: residents are wealthier, and better sources of revenue. Infrastructure and physical layout for installing wires and cellular transmission equipment is also more hospitable, and urban areas can serve as showcases of the nation's development. Only conscious efforts by government regulators can encourage the spread of universal nationwide service.

In contrast, the American private-ownership / government-regulated model showed that a private AT&T, willing to use its resources to invest in network expansion, could bring telecommunications tools to a vast segment of the population. Private management allocated resources effectively, though the FCC developed policies for universal service to ensure the company would serve citizens beyond lucrative urban areas. But government tolerance of the company's near monopoly tempered the risk to a significant degree.

How did China build its communications network over the past 150 years? The following sections highlight the patterns of growth in the sector, and the ways succeeding governments shaped the industry's development.

Imperial Building of China's Communications Lines, 1860s–1911

China was first exposed to telecommunications systems in the 1860s, as foreign colonial powers sought to connect telegraph lines in coastal cities such as Hong Kong, Shanghai, and Fuzhou. By this time, Hong Kong was already a British colony, and other cities were coming under foreign control as treaty ports. As Erik Baark's fascinating and comprehensive study of this period illustrated, Chinese national and local officials resisted allowing foreign traders to deploy telegraph lines in these areas.[47]

Chinese imperial officials, in a reaction similar to that of the contemporary Communist government, worried about the effect telecommunications would have on the Qing dynasty-controlled society. The government's office of foreign affairs (or *Zongli Yamen*) was concerned that telegraph lines would 'help foreigners spreading information undermining the authority of the Qing government'.[48] One official noted that the ability to spread information rapidly would allow the 'dissemination of unfounded rumours in newspapers, frightening people with [news about what has been] seen and heard'.[49]

A British businessman's attempt in 1865 to connect telegraph poles from Shanghai to the nearby coastal town of Wusong failed when the local population pulled down the poles, fearing they would disturb the area's natural harmony, or 'feng shui'. Farmers feared bad luck or illness would result if the lines trespassed ancestors' graves or crossed their fields.[50] In 1870, however, the Chinese agreed to allow the laying of an undersea cable from the British colony of Hong Kong to Shanghai. The Danish Great Northern Telegraph Company took the lead in connecting the line, even as it clashed with the Chinese in attempts to extend it inland. But by late 1871, the flow of telegrams, sent by both foreigners and Chinese, had begun on a steady pace.[51]

By the mid-1870s, the Chinese government had recognized the utility of the communications network, and even encouraged construction of a line from coastal Fujian province to the island of Taiwan for strategic reasons.[52] In the end, this line was not built at the time, but the Great Northern company did proceed to build a line within Fujian (which the Chinese government quickly took over). The Chinese themselves used the Danish company's materials to build short lines in Taiwan.[53] By the 1880s, the Chinese, with some help from Great Northern, were building extensive landlines connecting major cities such as Shanghai, Tianjin, Beijing, and even across to Taiwan.[54]

As telegraph connections grew, the government moved to establish central control over the lines. The Imperial Telegraph Administration (ITA) was established in late 1881 as a *guandu shangban* (official supervision, merchant-managed) private company.[55] The merchant shareholders running the company could keep profits, while the government retained authority over the enterprise, and could use the network to deliver military messages alongside private commercial communications. In fact, most of the managers running the company were apparently from government ranks, and were not really merchants. The company's chief, Sheng Xuanhuai, and his relatives and government associates seem to have been the primary shareholders. One commentary at the time noted that all the managers and assistant managers 'are officials of the ordinary type and know nothing about telegraphs'.[56]

Perhaps because of this quasi-private ownership structure, the ITA was given monopolistic control over telegraph communications. The expansion of the network to the more developed coastal cities of Suzhou, Hangzhou, and on to Guangzhou in 1884 was facilitated by government exemption of import duties on materials to build the line. Provincial governments contributed money to connect the western provinces of Yunnan and Guizhou. The government also provided security protection for the completed telegraph lines.

By the turn of the century, China had about 14,000 miles of telegraph lines under the control of the ITA, and another 20,000 miles of local lines managed by the provinces. The newly deployed telephone network (discussed below) also came under ITA control in 1900.[57] But in 1902, the government decided to nationalize the ITA, and in 1909, completed the process of buying out the private shareholder interest. The government took this step to expand lines to areas not commercially viable, to provide better line maintenance, and to keep usage costs low. As a quasi-private monopolistic company preceding full nationalization, the ITA had few incentives to improve the lines or communications services.[58]

To replace the ITA administration of the communications network, the Qing government established the Ministry of Posts and Communications in late 1906. The ministry combined the telegraph industry with railroads, shipping, and the postal system. The ministry was the only Imperial one to have stable annual revenue, as it could count on profits from the country's rail network.[59] Sheng Xuanhuai and a rival, Tang Shaoyi, alternated as head of the ministry in its last years before the end of the imperial rule in 1911.

Telephone lines soon followed on the heels of telegraph communication. British merchants were the first to set up telephone lines, in Shanghai in 1881.[60] In 1891, Sheng Xuanhuai requested the imperial government to allow the building of a telephone network under his administration. The first Chinese telephone system was established in Nanjing in 1900. In 1910, the year before the end of the Qing dynasty, the Chinese network had more than 7,000 phone subscribers. Foreign companies also had the right to establish phone networks in their settlement areas in China, and had a similar number of subscribers, mainly run by the Japanese and Russians in Manchuria. Shops owned about 40 percent of Chinese telephones, government offices had 26 percent, and only 20 percent were in private hands.

After some false starts, the Chinese Imperial government had moved to expand its communications network across the nation. The ITA and local governments did succeed in creating a rather broad system of telegraph lines, and even the telephone network saw progress. However, by the end of the Qing dynasty, the monopolization of telecommunications under quasi-private control had led to general stagnation, and profits from the industry found themselves mainly in the hands of those officials who controlled the communications systems.

Development of Communications under the Republic of China (1911–1949)

China deteriorated into rule by regional military rulers (or warlords) in the first decade after 1910, and civil war damaged many of the nation's telephone and telegraph lines.[61] By 1927, the new Nationalist (or 'Kuomintang'—KMT) government under the control of Chiang Kai-shek had taken control of most of eastern China (except for Manchuria in the northeast, which was ruled by Japanese puppet officials). The new government, now with a capital in Nanjing, set out to reconstruct some of the damaged infrastructure.

Under the new regime, the Ministry of Communications continued central government ownership and regulation of the industry, in coordination with provincial telephone administrations in each major province.[62] From 1927 to 1934, the KMT government repaired 4,000 miles of telegraph lines, and built 1,800 miles of new lines.[63] The Chinese government also took back control of international telegraph lines in 1930, following many decades of foreign company control.[64]

As for telephones, by the mid-1930s, many major coastal cities had local phone service, and there were long-distance connections between the capital Nanjing and cities such as Shanghai, Tianjin, and Hankou, as well as between Beiping (later Beijing) and Tianjin. The Nationalists worked to expand the network province by province, with Jiangsu, Zhejiang, Anhui, Henan, and five other provinces connected utilizing the existing telegraph poles and space alongside roadways.[65] By 1936, the nine-province network was essentially complete, and the country had some 47,000 kilometers (29,000 miles) of long-distance phone lines.[66]

Much of the telephone expansion was controlled at the local level. For example, each provincial telecommunications administration moved to establish its own long-distance telephone department. Zhejiang province used its own resources to build out its local network, and paid part of its revenues to the central ministry as license fees. Guangdong province developed plans to link its lines with neighboring Guangxi.[67]

The Japanese occupied much of China from 1937 to 1945, and took control of the nation's communications network. To strengthen their rule, the Japanese expanded and improved China's system, and as of 1946, the country had about 59,000 kilometers (37,000 miles) of long-distance phone lines, and cities had some 108,000 phone subscribers.[68] The number of subscribers grew even during the country's civil war years of 1946–9, and by the beginning of the Communist era, there were a total of 218,000 telephone users in the nation's urban areas.[69]

The Republican era saw continued government control of the network, with central–local cooperative efforts to expand the reach of the system. Many of these decades were plagued by warfare, so it is hard to say whether the Nationalist model would have succeeded in providing the nation with a modern and widely accessible communications system. The new Communist government would have a whole new set of motivations for expansion and control of China's telephone network.

Telecommunications during the Reign of Mao Zedong (1949–1976)

On 1 November 1949, the new Communist Chinese government led by Mao Zedong officially founded its own Ministry of Posts and Telecommunications (*youdianbu*), or MPT.[70] The new ministry effectively took control of functions held by the Nationalist government's Ministry of Communications. It became the country's state-owned monopolistic posts and telecommunications administration, or 'PTA'. It was both the regulator of telecommunications services, as well as owner and operator. But other industrial branches of the government manufactured telecommunications equipment (as is discussed later in Chapter 5).

The new regime inherited some useful infrastructure. Overall, there were some 310,000 phone lines, with 200,000 of these automated, and about one-third of the nation's counties had local phone service. Still, fourteen provinces, cities, and autonomous regions lacked any kind of automated phone service. Long-distance communication depended on wireless short wave, as there was no nationwide network. The vast majority of villages had no communications infrastructure—the few lines that did exist in the countryside were often for use by regional military leaders.[71] (Chapter 7 focuses on the urban–rural divide.) For a now united nation with some 550 million people, the number of phone connections was still small, with only one line for every 1,800 people.

Though the number of telephones was limited in the early years of Communism, there were still methods of 'mass communication' in the broader sense of information transfer.[72] Among these were radio, film, newspapers, and *dazibao* (big character posters).[73] Of course, most of these forms of transmitting information were one-way, mainly flowing from government propagandists to the masses. For political purposes, however, these methods, along with the introduction and rapid spread of television in the 1970s and 1980s, were sufficient to convey information—the lack of telephones was no hindrance to keeping people informed of the news and policy statements the authorities deemed vital.

Still, the government did put effort into extending the telecommunications network, and the number of subscribers grew, as Figure 2.1 indicates. In 1956, Mao himself reportedly indicated his support for building microwave and cable communication trunk lines, and asserted that the country should strive to catch up with advanced world standards.[74] From 1950 to 1960, the total fixed assets of the telecommunications sector grew at an annual clip of 9.1 percent. Though this was solid progress, it trailed

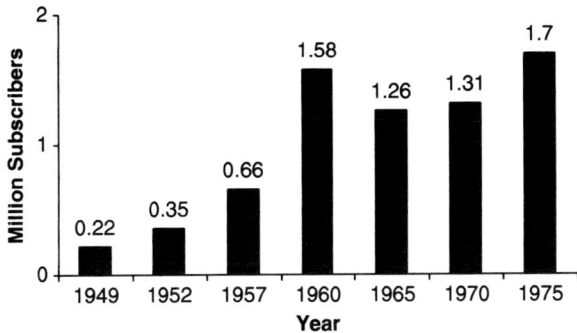

Fig. 2.1. Number of telephone subscribers, 1949–1975
Source: Statistical Yearbook of China, 1981 (English edn.), Hong Kong: Economic Information & Agency, 1982: 295.

expansion of the nation's agricultural and industrial production (which grew at a 15.7 percent annual rate) as well as national income (at 11.7 percent).[75] Moreover, this progress had some unseemly support: according to one report, as of 1955, there were as many as one million Chinese working as forced laborers on telecommunications projects.[76]

In addition to regulation and operation of the communications network, the MPT also oversaw much of the production of telecommunications equipment. Chapter 5 has more detail on the growth and development of equipment manufacture. In brief, however, equipment production mirrored Soviet-style state control of most other industries: foreign and private ownership of factories ceased, and the state took control of nearly all means of production. The ministry guided industries by setting quotas in consultation with a provincial or municipal telecommunications authority office. Production plans would then be reviewed and audited by the national State Planning Commission (SPC) before a final version was sent to the equipment-producing factory.[77] The more important policy decisions were made at higher-level commissions, such as the State Economic Commission (SEC) or SPC, or even the cabinet-level State Council or Communist Party Central Committee.[78]

Essentially, political control of the telecommunications sector was top-down, with provincial and lower-level officials having less say in the industry's direction. At the county level, for example, the Post and Telecommunications bureau and its Telephone–Telegraph group were much more tightly linked to central authority control after the 1949 revolution.[79] During the Mao years, 'provincial and other local organs had no inviolable autonomy...the powers they exercised were only those delegated from the center'.[80] In the Chinese lexicon, vertical links (*tiao*, or 'lines') dominated over horizontal or regional ties (*kuai*, or 'pieces') for building and managing the communications system.[81] The desire to coordinate the growth and management of a nationwide telecommunications network contributed to the special need for central authority in the sector.

Because of the importance of vertical control, the MPT was hampered over the 1950s and 1960s by lack of a powerful advocate at its helm. Minister Zhu Xuefan governed the MPT from its founding in 1949 to his purge during the nation's Cultural Revolution turmoil in 1967. Though Zhu had studied law in Shanghai and at Harvard University in the 1930s, his grounding was solidly in the postal union sector, rather than in telecommunications. He also had been closely allied with the KMT's gangster-influenced labor movement, and only moved to the Communist camp in 1948.[82]

China's 'Great Leap Forward' campaign of 1957–9 included many irrational moves toward more local control of industrial production and fiscal resources. In November 1957, for example, the central government mandated that local authorities should be given 20 percent of enterprise profits. Significantly, the Ministry of Posts and Telecommunications was one of a handful of ministries exempted from this decree, and all profits continued to go to the central PTA authorities.[83] Still, the ministry suffered from poor construction methods and inaccurate counting of assets, which likely accounts for the inflated number of phone lines noted in Figure 2.1. According to one report, 'one of the reasons cited for the vast confusion of the Great Leap Forward was the excessive use and tying-up of the telephone system by bureaucrats who had found a new toy'.[84] (Chapter 7 has more detail on telecommunications construction in rural areas during the Great Leap Forward.)

Following the Great Leap Forward, more pragmatic Chinese leaders, including Liu Shaoqi and Deng Xiaoping, took a greater role in political matters and economic planning. The years 1961 to 1965 also saw moves toward acquiring Western technology, as the Soviet Union had withdrawn its advisors in these years. Chinese telecommunications experts visited facilities in Western Europe and Japan to learn about the latest equipment for telephone transmission, telex, and fax services. The PRC ordered complete plants to produce vital electronic components.[85]

These efforts produced few results, however, as the Cultural Revolution of 1966–76 brought chaos to nearly all parts of China's political and economic system, and ended much of the contact with developed nations. The Ministry of Posts and Telecommunications was also a victim of the movement. In late August 1967, the MPT came under military control, likely to protect the facilities from radical Red Guard youth activities that could undermine Maoist control over the network. But minister Zhu Xuefan disappeared that same year, and the ministry itself was abolished at the beginning of 1970. It was replaced by the General Telecommunications Administration, a unit under a newly constituted Ministry of Transportation. A former general and MPT vice-minister, Zhong Fuxiang, led the administration. As the Cultural Revolution subsided in the early 1970s, the MPT was restored in 1973, and Zhong continued as minister until he was replaced in 1978.[86]

As the rest of the economy slowed, so did investment in telecommunications fixed assets, which grew at an annual rate of only 4.6 percent from 1960 to 1970.[87] In 1960, China had one phone for every 410 people; by the end of the decade, as the population continued to grow, the number had shrunk to one for every 630 people. A severe frost in January 1969 caused severe disruption of the network in nearly all of the northern and central parts of the country—this event signaled to the regime that it could no longer continue to neglect the development of the communications system.

Telecommunications growth was relatively slow in these decades partly because of the Communist Chinese system of enclosed work units, or *danwei*.[88] Restrictions on place of work and movement, in particular for urban residents, meant relatively little need for communication outside of a set workplace. Manufacturing was often done on a self-reliance basis, and supply companies were located close to the assembly site. In rural areas, close-knit family ties in villages also lessened the need for voice telecommunication. Finally, China's international isolation under Mao meant phone calls to places outside of the country were both less common and less necessary than for other developing nations.

One foreign visitor to China in 1966 noted there were many alternatives to telephone use for sending information, at least for city dwellers.[89] Enterprises could send messengers by foot or bike, and wait for responses to come the same day. The telegraph system served as another alternative for sending long-distance messages. In my own experience in China in the early and mid-1980s, telegraph messages were a convenient and relatively inexpensive alternative to a visit to a long-distance telephone office for sending a message to a Chinese city or town outside of Beijing.

Later chapters in this book will have greater detail on developments during the Mao years. Overall, however, we can note several key elements of telecommunications development during the first decades of the PRC. First, the new regime worked to restore service disrupted by the years of Japanese control and civil war, and to extend telecommunications to many new parts of the nation. Though overall investment was not a significant part of the nation's budget, there was steady expansion of telephone access in the 1950s and into the early part of the 1960s. The Cultural Revolution of the mid-1960s to mid-1970s led to a slowing of the industry's development.

In a larger sense, telecommunications did not figure as a priority for the new regime. Mao Zedong and his allies put greater emphasis in these years on consolidating political control and sorting out social and class conflict issues. Where industry was considered, heavy industrial sectors received greatest attention. During the Cultural Revolution, telecommunications was only vital for issues of political and security communications reasons; there was little incentive to expand the network for commercial functions or to meet societal demand.

The leaders who followed Mao saw development of telecommunications in a different light, and instituted policies that shaped the sector for the next thirty years. The following sections highlight the main policies and actors that laid the groundwork for events of these recent decades.

Politics of Telecommunications in the Early Reform Era, 1976–1993

As was typical in the early days after Mao's death in September 1976, the new leaders used declarations from the late party chairman to justify steps to transform the telecommunications sector. In 1977, the official *People's Daily* newspaper cited some of Mao's statements from 1956 and the mid-1960s in support of expanding the network. In a departure from the Cultural Revolution rhetoric on political struggle, the article suggested China should aim for 'all-round mechanization, automation and adoption of electronic techniques throughout the field [of communications] by 1985, and to overtake advanced world levels before the end of the century'.[90] Mao's immediate successor as Communist Party chairman, Hua Guofeng, declared that those in the industry should 'work hard to achieve the modernization of posts and telecommunications'.[91]

Several factors inspired greater attention to telecommunications development.[92] Mao's death and the subsequent turn toward 'market socialism' meant a greater need for rational and coordinated supply systems. Worker mobility also increased as the *danwei* structure began to weaken. Rural migration to the cities grew, and foreign investors came to attract low-wage labor to new factories in many parts of the country. These changes required a more developed system for both business and personal communication.

But change came slowly. As noted above, the Cultural Revolution MPT leader Zhong Fuxiang retained his post until 1978, when Deng Xiaoping was finally able to wrest policy control from his rival Hua Guofeng. But two new MPT ministers, Wang Zigang (in office from 1978 to 1981) and Wen Minsheng (minister from 1981 to 1984) were also largely ineffective at moving forward telecommunications development policies. Wang had been one of Zhu Xuefan's vice ministers from the 1950s to the Cultural Revolution. Though he seems to have had more of a technical background than Zhu, there is little evidence he helped move the ministry forward in the early years of reform. Wen's background was mainly in provincial political leadership (he had been governor or party leader of Guangdong, Henan, and Heilongjiang provinces during the 1950s, 1960s, and early post-Mao years).[93]

Political change in a larger sense began in 1980, when the government called for a smaller role for the Communist party in directing the state-owned industrial sector, and greater powers for those directly involved in production. Deng Xiaoping advocated reforms to 'take the party committee out of day-to-day affairs [at the enterprise level], and allow it to concentrate on political and ideological work and organizational supervision'.[94] As noted below, succeeding telecommunications ministerial leaders were chosen more for technical expertise than for ideological loyalty.

Signs of reform in the telecommunications arena also began to appear in 1980. Until that year, revenue from telecommunications operations were turned over to the central government, and the industry received a set percentage of the budget (no matter what the receipts) for development. But in a new policy the government listed posts and telecommunications as a preferential construction item, and allowed the industry to use funds derived from its operations at national as well as local and collective levels to supplement its development. By 1986, some 50 percent of the industry's revenue came from this 'self-raised' capital.[95]

Political attention to telecommunications began to focus further in the mid-1980s, with 1984 a key turning point year. Yang Taifang, who had a degree in electrical engineering from China's Zhongshan University, became the MPT minister. He joined a growing cadre of top government officials who had both completed college, and focused on the apolitical field of engineering. Yang had been active in telecommunications administration since he was 26 years old as, in 1953, he was named a section chief at the Guangdong province posts and telecommunications office.[96] He rose through the MPT ranks, and became a vice minister in 1982. Aged 57 at the time of his appointment to the MPT top post, Yang was part of a new breed of technocratic leadership.

The same year Yang was appointed, top central-government leaders took steps to show their commitment to strengthening the sector. The Communist party central committee and the national State Council issued important decrees signifying the importance of the sector. In the so-called 'Two Six-Point Targets'(*liang ge liu tiao zhibiao*), the government put the improvement of posts and telecommunications at the same level as the more strategically vital areas of energy and transportation construction. The government noted that China should not necessarily imitate foreign telecommunications management structures, and should encourage a 'double administrative system' in which local telecommunications authorities played a role under the central MPT leadership.[97] The year also saw then vice premier Li Peng taking control of another top-level economic body, the Leading Group for the Revitalization of the Electronics Industry. Under this group's guidance, telecommunication was put at a higher place on the government's agenda.[98]

The government announced telecommunications targets at the same time it made its calls for developing the industry. The 1984 decree set a long-range goal of increasing telecommunications capability sixfold from 1980 to the year 2000.[99] The following year, the MPT submitted a 15-year plan for telecommunications development to the State Planning Commission. The goal in this plan called for a teledensity (lines per 100 people) of 2.8, for about 30 million lines, by the year 2000, up from about 6 million lines and teledensity of 0.6 in 1985.[100]

Over the 1980s and into the next decade, central government leaders issued policies and directives to ensure the industry would see rapid growth. One of the key policies to inject funds into telecommunications, which came to be known as the 'Three 90 percents', also played a major role in developing telecommunications. In 1982, the State Council stipulated that provincial telecommunications companies need only give 10 percent of their taxable profits to the central government, and could reinvest the other 90 percent. They could also keep 90 percent of their non-trade foreign currency revenue, mainly derived from international telephone traffic. In 1986, the government added a third policy, allowing the companies to avoid repaying 90 percent of the principal and interest on central government loans.[101]

The State Council issued other key directives in these early years to spur communications growth. In 1989, for example, Directive 56 limited foreign participation in switching equipment manufacture to three companies, one each from France, Germany, and Japan. Chapter 5 focuses on the equipment sector in detail, but we note here that the directive both reduced duplication of effort for equipment deployment while it added competitive market forces that lowered prices for building out the communications network.

Another important directive came in 1993, and also inspired greater competition, this time in the area of domestic communications service provision. Directive 178 created the first new telephone company of the Communist era, known as Liantong, or China Unicom. In the wake of the creation of China Unicom, mobile phone service expanded at a rapid clip. Chapter 3 gives more detail on this case of the creation of a new mobile phone rival to the state monopolistic carrier.

Structural reorganization of the industry was also conducive to its development. Before 1990, the provincial offices took direct orders from the MPT. Following that year, the provincial branches received autonomy in getting equipment, making investments, planning, and accounting for their performance.[102]

These policies taken together were quite successful, as the results seen in Table 2.1 and Figure 2.2 indicate. Table 2.1 shows the growth in telecommunications investment vs overall GDP growth from the beginning of the reform era to 2006. As the table indicates, the fastest growth in telecommunications investment (relative to the expansion of the economy as a whole) came in the years 1985 to 1994. While the nation's GDP grew an average annual pace of about 20 percent over this period, investment in telecommunications rose at nearly triple the rate, with an annual increase of some 55 percent. In total numbers, GDP grew a little over fivefold, while telecommunications investment surged nearly 52-fold.

Figure 2.2 shows that the government's 1985 goals to expand communications infrastructure were achieved far ahead of schedule. The aim of 30 million lines was met at the end of 1993, and by 1996 there were already about 93 million lines. The goal for 2000 was changed to 120 million lines, and a teledensity of between 8 and 9.[103] The actual result for the year 2000 was 178 million phone lines, and 145 million actual fixed-line subscribers. In what planners in 1985 likely failed to imagine, there were an additional 85 million mobile phone subscriptions that year. Adding these numbers together, there were some 230 million phone accounts in 2000, for a teledensity of subscribers per 100 people of about 18.[104]

China's Telecommunications History

Table 2.1. Fixed-asset investment in telecommunications as percent of gross domestic product (GDP), 1978–2006, (selected years)

	Investment (in billion yuan)	GDP (in billion yuan)	Investment as percent of GDP
1978	0.255	365	0.07
1985	1.34	902	0.15
1987	2.14	1,206	0.18
1989	4.01	1,699	0.24
1991	6.89	2,178	0.32
1992	13.7	2,692	0.51
1993	35.3	3,533	1.00
1994	69.4	4,820	1.44
1995	85.6	6,079	1.41
1996	91.1	7,117	1.28
1997	105.6	7,897	1.34
1998	150.1	8,440	1.78
1999	160.5	8,968	1.79
2000	222.4	9,921	2.24
2001	255.3	10,966	2.33
2002	207.3	12,033	1.72
2003	221.8	13,582	1.63
2004	219.9	15,988	1.38
2005	209.8	18,387	1.14
2006	218.7	21,087	1.04

Sources: *Zhongguo jiaotong nianjian* (China Communications Yearbook), Beijing: Zhongguo jiaotong nianjian she, 1990: 560 (for 1978 telecommunications investment); 1991: 621 (for 1985–9 telecommunications investment); var. years for 1985–2005 investment; Xinhua Economic News Service, 23 Jan. 2007 (for 2006 investment); and *Zhongguo tongji nianjian* (China Statistical Yearbook), Beijing: Zhongguo tongji chubanshe, 2007: 57 (for GDP figures).

As of 2005, there were about 57 subscribers per 100 people, putting China roughly on a par with levels seen in advanced nations such as Japan and Australia in the early 1990s.[105] By mid-2007, the number had risen to some 66 per 100 people.[106] Table 2.2 puts China's more recent rapid progress in a comparative context—the following chapter gives more detail on reasons for growth in the 2000s. However, as Chapter 7 will discuss, many of China's telecommunications resources were concentrated in relatively wealthy urban areas, leaving poor urban and rural members of society without communications access.

Overall national growth was so quick in the early 1990s that by late 1995 the 'Three 90 percents' policy was repealed, as it was seen as no longer necessary.[107] In succeeding years, about one-third of the revenue for telecommunications expansion came from telephone installation fees, which in 1997 averaged about 2,000 to 3,000 yuan (about $240 to $360) per line. Another third came from depreciation, which, at a favorable rate of 16 percent and asset value of 500 billion yuan ($60 billion), meant the sector could utilize tens of billions of yuan per year. The final third of funds for expansion came from domestic and foreign loans. The MPT also received,

China's Telecommunications History

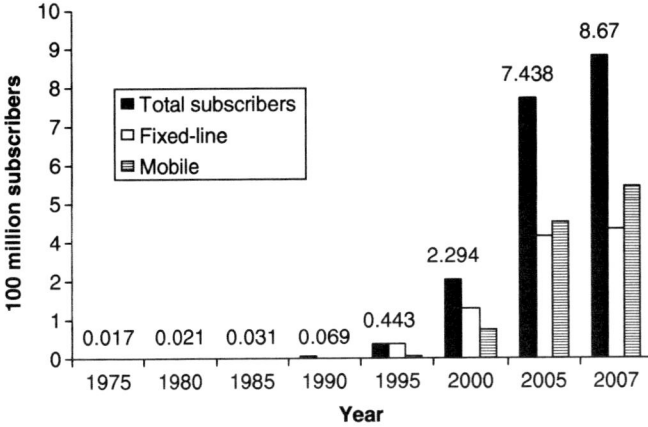

Fig. 2.2. Number of telephone subscribers, 1975–2007 (selected years)

Note: Figures for 2007 are for May of that year.

Sources: For 1975: *Statistical Yearbook of China, 1981*, (English edn.), Hong Kong: Economic Information & Agency, 1982: 295; for 1980–2000: *Zhongguo tongji nianjian* (China Statistical Yearbook), Beijing: Zhongguo tongji chubanshe, 2005: 584–5; for 2005: ibid. 2006, CD-ROM ver., chart 16–40; for 2007: Xinhua Economic News Service, 6 July 2007 report.

Table 2.2. Numbers of telephones and teledensity for China and other selected nations, 1999 and 2005

Nation	Total number of telephone subscribers (fixed-line and mobile) in 1999 (in millions)	Teledensity (subscribers per 100 inhabitants) in 1999	Total number of telephone subscribers (fixed line and mobile) in 2005 (in millions)	Teledensity (subscribers per 100 inhabitants) in 2005
Australia	16.1	84.94	28.5	141.60
China	**152.0**	**12.03**	**743.9**	**56.53**
India	28.4	2.83	140.3	12.72
Japan	118.9	93.88	154.5	120.65
South Korea	49.1	107.4	62.1	128.56
Spain	31.5	78.3	60.6	142.04
Thailand	7.6	12.42	38.2	59.43
United States	275.5	98.75	388.4	130.23
Vietnam	2.4	3.14	25.4	30.20

Source: International Telecommunications Union (ITU) website, http://www.itu.int/ITU-D/icteye/Indicators/Indicators.aspx#.

as of 1997, some 100 million yuan ($12 million) per year to cover medical and other institutional expenses.[108]

Later chapters and this book's conclusion will explore further policies that guided the sector in the early years of the new millennium. But the key political choices that shaped the industry in its critical years came in the 1980s and 1990s. As Table 2.1 and later chapters will show, these early decisions by government leaders were vital to the rapid development of the communications network into the next decade.

MPT Leadership after 1993

Though Yang Taifang had presided over the beginning of telecommunications expansion in the 1980s and early 1990s, the fruits of his ministry's efforts emerged after his departure from the post. Yang had served a full eight years, and premier Li Peng's new five-year term, beginning in 1993, brought an opportunity to rearrange the national State Council. The MPT leader for the rest of the decade would oversee the introduction and regulation of key new technologies, including the Internet, mobile phones, and the convergence of data and voice networks.

This new MPT leader was Wu Jichuan, a man who carried on the trend of being an administrator with a college degree in engineering.[109] Wu, born in 1937 in Hunan province, had studied wire communications at the Beijing Institute of Posts and Telecommunications. He joined the Communist Party only in 1960, a year after his college graduation. Following the Cultural Revolution, Wu joined the MPT in 1978, first as a division chief in the bureau of materials, then as deputy planning director. In 1984, at the age of 47, Wu became an MPT vice minister under Yang Taifang.

Wu's star continued to rise at the beginning of the 1990s, when he briefly served as one of three deputy Communist Party Secretaries in Henan province. After more than two years of regional administrative experience, he returned to the MPT as one of the country's youngest ministers (aged 55) in March 1993.

Following chapters will give further detail on policies pursued during Wu's reign at the MPT and successor Ministry of Information Industry (MII). Overall, however, Wu presented a figure of great authority at the ministry, and came to be dubbed in 1990 by the influential *Economist* magazine as 'the most powerful man in what is likely to become the world's biggest telecoms market'.[110] Wu was known for his opposition to opening China's telecommunications and Internet markets to foreign companies (discussed in more detail in Chapters 3 and 4), and struggled with China's prime minister, Zhu Rongji, over this issue.[111] In 1999, Zhu reportedly tried to fire Wu, arguing his ministry's control over the communications network was too restrictive—but Wu was able to survive until the end of Zhu's term in March 2003.[112] By this time, the telecommunications revolution had already made a profound mark on the Chinese landscape.

Conclusion

Even before the rise to power of the Communist Party in 1949, China had chosen the path of government monopolistic control over its communications sector. As early as the nineteenth century, imperial administrators recognized the strategic and national security concerns related to information distribution. The government was also keen to retain the revenues generated by the new technology.

War and revolution marred the first part of twentieth-century China's development, and telecommunications tools were slow to spread outside of areas controlled by foreign powers. Nationalist government efforts in the 1930s to expand the network were short-circuited by the Japanese invasion and subsequent civil war. But even the KMT government did not take steps to end the public monopoly.

The first few decades of Communist rule also saw China staying with the typical model of state-controlled monopoly. Like early Britain, France, and many developing countries, Mao-era China failed to commit the necessary resources to expand the network beyond a limited number of the nation's elite government officials and citizens. By the early 1950s, there was no private sector in existence that could have even tried to challenge bureaucratic control of the network.

As this chapter has indicated, new government policies in the late 1980s and early 1990s led to a dramatic expansion of the network, even as it remained in public hands. The government moved to put technocrats with engineering backgrounds in positions of authority, rewrote rules to allow state corporations to retain more of their funds, and oversaw a dramatic increase in the budget for the telecommunications authorities. The officials used the funds effectively, investing to expand the network at a speed unseen in any developed or developing nation to date.

Unlike other developing nations, such as the Latin America cases cited at the beginning of this chapter, China has been cautious about allowing foreign companies a significant role in the telecommunications sector.

The issue of foreign investment in phone and Internet operations is explored in more depth in Chapters 3 and 4, and Chapter 5 considers the equipment-building part of the industry. Each of these chapters will indicate, however, that even here, the Chinese were able to benefit by holding off foreign investment until their own resources had reached a significant degree of technical maturity.

The first two decades of reform in China showed that a well-crafted set of telecommunications policies could lead to dramatic improvement in communications abilities. The government avoided the potential problem of stagnation that a state-run monopoly, insulated from both domestic and foreign competition, could face.

By the mid-1990s, however, government leaders realized that the forces of competition could add a key element to the sector's continued growth. The 1984 breakup of AT&T and the resulting new leap in the American telecommunications industry showed the Chinese that a new model could help maintain growth in the industry. The following chapter shows how the Chinese government shifted gears, and allowed competition to herald in the new telecommunications technology of mobile phones.

3
Telecommunications Competition in the 1990s and 2000s

In theories of industrial policy, a key problem is ensuring competition among enterprises, many or all of which may be supported or actually owned by the state. When foreign competition is lacking, one should expect even more that companies could become lax.

Yet China's telecommunications service sector, after the mid-1990s, became a shining example of effective competition among state-owned enterprises, one resulting in improved service, lower prices, and generally greater access for the country's citizens. This chapter shows that the added element of rivalry between bureaucratic entities was a key ingredient for fostering competition. This competition was further fueled by the ability of state enterprises to attract foreign funding in the form of both direct investment and overseas stock listing.

The chapter focuses first on the case of China Unicom, a company the central government formed specifically to foster competition. As the chapter will show, the government expected competition would result in improved services and lower costs. Despite some initial reluctance of the Ministry of Posts and Telecommunications (MPT) to allow corporate rivalry, missteps by the new Unicom company, and a somewhat misguided effort to gain foreign capital and technology, the results, as the chapter will show, were extremely positive.

Moves further to develop competition in the early 2000s had mixed results. Some new companies emerged as potentially powerful participants in the emerging areas of mobile telephony and data transmission. Others, however, seemed to be orphaned at their birth, with little chance to grow into rivals of the established corporations.

Competition in the 1990s and 2000s

Before considering the China case, the chapter briefly assesses patterns in other developed nations, and the effects competition has had on their telecommunications markets.

Patterns of Telecommunications Competition

Since the break-up of the private American carrier AT&T in 1984, many countries have recognized the value of competition for advancing the telecommunications sector. A key factor differentiating telecommunications in different nations is the degree of ownership by the national government and, in some cases, municipal government as well.

As the last chapter discussed, in the American case, one private corporation, AT&T, dominated the sector for some seventy years. The birth of the regional Bell companies in the 1980s and the new equipment producer Lucent Technologies (the former Western Electric Manufacturing Company) in 1996 led to intense competition to provide better and cheaper services and telephone equipment, and contributed in the 1990s and early 2000s to the widespread use of mobile phones as well as broad-based access to data networks.[1]

The lessons of competition were not lost on other nations, many of which moved to emulate the American example. In Europe, where government-owned telecommunications corporations were dominant, there was a dual move to privatization and dilution of monopolies. The United Kingdom privatized a majority share of the public monopoly British Telecom (BT) in the early 1980s, and, by the early 1990s, competition with rival Mercury Communications and others contributed to a rapid fall in areas such as mobile call prices.[2] By 2004, BT had 269 licensed rivals, and ranked second in size to competitor Vodaphone.[3] France was slow to end the monopoly of state carrier France Telecom, but European integration forced the country to allow new rivals from 1998. In 2004, the government still owned 53 percent of the company, but planned to become a minority owner.[4] That year, deregulation allowed French citizens to enjoy widespread and cheap broadband voice and data communications access.[5] In both cases, the trend was from public to private ownership, and from monopoly to competitive market; the result was lower consumer costs, broader access, and a wider array of services.

Japan also moved to break up its monopoly NTT in the 1980s and 1990s, but had mixed results. The company was partially privatized in 1985, but

Competition in the 1990s and 2000s

the government continued to hold a majority stake into the late 1990s. In July 1999, NTT was divided into two regional phone companies (NTT-East and NTT-West) and one long-distance company. Mobile service was assigned to NTT DoCoMo. All of these companies continued to perform under the guidance of an umbrella holding company.[6] But competition did emerge in the mid-2000s from private companies such as Softbank and KDDI, which set up rival networks and forced the NTT group companies to cut rates to remain competitive.

As Chapter 2 discussed, developing nations were generally reluctant to yield government monopoly control over telecommunications service and equipment manufacture, and foreign companies were kept at arm's length. In many African countries, as noted earlier, national governments continued to dominate all aspects of communications up to the 1990s.[7] The previously cited examples of Kenya and India, however, indicated that even developing countries made moves toward relaxing monopoly control and allowing private-sector competition in the industry.

On a broader scale, we can ask: is a telecommunications monopoly necessary or even beneficial for developed or developing nations? The telecommunications theoretician Eli Noam explored some of the key points.[8] He asserted a public monopoly is not necessary for universal service, for example, as government subsidies can make up for the lack of attractive rural markets (as discussed in this book's Chapter 7). Even in the American private monopoly case, small private companies serviced rural parts of the US.

Other arguments in favor of long-term monopoly control are weak. Equipment compatibility for new, competing companies is not normally a problem, as international bodies set technical standards that prevent problems from arising. The assertion that competition will lead to duplication of effort and excess capacity may be true, but this is common in many industries, and the result (as the China case will show) is often lower prices for services and benefits for the consumer.[9]

In the end, as the following sections chronicle, the Chinese government recognized the utility of competition within the telecommunications service industry. However, the desire to maintain control of the strategically key sector meant that telecommunications would remain in government hands, with foreign companies and even Chinese private investors barred from participation.

Breaking the MPT's Monopoly: The Early Years of China Unicom

As Chapter 2 discussed, the MPT took over control of both telecommunications operations and regulation in 1949, and managed the sector under public monopoly control throughout the Mao Zedong years. One of the foundations of Mao's rule was the elimination of 'bourgeois' capitalist private ownership, so there was no chance that rival private telecommunications companies could appear. Furthermore, there was little thought that even state-owned competition could be of benefit to consumers, as prices for most commodities remained fixed throughout most of the Mao years. In any event, the vast majority of citizens lacked the means to subscribe to communications services, so competitive pricing would have been essentially meaningless.

The MPT's monopoly on telecommunications services lasted during the early years of the reform period. As Chapter 2 discussed, the government saw little progress at the beginning of the 1980s, and was open to examining other nations' models to see how to encourage faster growth. The 1984 decision to break up AT&T in the United States encouraged the Chinese telecommunications authorities to think about the role competition could play in fostering an efficient telecommunications sector.

While considering moves to break the MPT's monopoly on communications services, one cardinal rule for restructuring the sector and introducing competition would be ensuring the entire communications infrastructure stayed in government hands. A paramount concern was security—communications infrastructure in the hands of private citizens or foreign corporations could threaten the government's control over the flow of information. Dissident voices could potentially use an open network to challenge the regime, and foreign participation could pose a threat to national security or civil stability.

Another factor behind keeping the industry state-owned was control of revenue. Foreign companies coming to China, with vastly more experience in owning and operating communications systems, stood to sap the potentially lucrative sector of great future profits. As in many other sectors, such as automobiles, finance, and agriculture, the government sought to limit the ways foreign interests could participate in the industry. As we will see, foreign capital did work its way into the sector, though with less than optimal results for most of the participants.

As the move to introduce competition proceeded, the State Commission for Restructuring the Economy, then led by premier Li Peng, sent a team

Competition in the 1990s and 2000s

from the MPT, the Chinese Academy of Sciences, and other related groups to the US in 1988 to find how the Federal Communications Commission (FCC) managed the telecommunications landscape following the dismantling of AT&T.[10] The team came away with the idea that competition was helpful, and was also impressed with the importance of separating regulatory authority from corporate operations. In the China case, these powers were combined in the MPT.

The political restructuring following the June 1989 student demonstrations in Beijing took telecommunications reform off the political agenda, as all major changes were temporarily frozen. The MPT grasp on both operations and regulatory power, however, was again challenged in 1992, as the country began a new wave of economic and political change. Some international pressure to break the state monopoly on telecommunications came from the World Bank, a key source of revenue, which encouraged the broadening of market forces. The Bank also wanted to see the spread of telecommunications to rural areas in China.[11] As Chapter 7 will discuss, these were parts of the country that the MPT itself was less eager to develop.

Early calls for reform came from the Ministry of Electronics Industry (MEI), which had manufactured equipment for the industry but now wanted a piece of the operations pie. New minister Hu Qili, resurrected from a post-1989 purge of the Communist Party politburo, led the battle to forge a rival to MPT operations control. In addition to the revenue and prestige the MEI could win by becoming a player in telecommunications operations, Hu may have also seen a new company as a way to accelerate the country's advance into the information age.[12] In 1992, the MEI formally suggested forming a new telecommunications company. MEI allies included the Ministry of Railways (MOR) and Ministry of Electric Power (MOP), both of which had their own separate telecommunications networks and, at the time, excess capacity.[13]

The idea that ministries outside of MPT control should found their own competing telecommunications company was a fundamental breakthrough. The system would still stay out of private or foreign hands. The state would still be the owner of the networks. But the bureaucratic rivalry inherent in the competition for resources and prestige between the different ministries would be a powerful engine to drive a new telecommunications company to fight full throttle to break the telecommunications monopoly. The alliance of companies with rival ministries could be exactly the right tool to introduce the intra-state corporate competition that would lead the communications network to new heights of efficiency and consumer responsibility.

Competition in the 1990s and 2000s

In early 1993, Hu courted top-level leaders to win their support for the new company. Premier Li Peng apparently favored the project as it was slated to include the energy industry, a sector with which he and his family had close ties. Vice Premiers Zhu Rongji and Zou Jiahua also supported the formation of a new company.[14]

In December 1993, the national State Council issued its Directive 178, formally approving the creation of a new telecommunications carrier, known in Chinese as *Liantong*, or China Unicom.[15] In founding Unicom, the MEI and ministries of Electric Power Industry and Railways were joined by twelve other partners officially to open Unicom in July 1994. The MEI, Power, and Rail ministries contributed 100 million yuan (about $12 million) each for a 7.5 percent ownership share, while the remaining thirteen partners each put in 80 million yuan ($9.6 million) and held about 6 percent apiece.[16] The total investment came to about 1.3 billion yuan (some $160 million).

The Unicom members together represented a potentially powerful array of stakeholders, just the kind that could challenge the powerful MPT monopoly.[17] Table 3.1 lists the shareholders and their ownership shares.

At its founding, the company was also granted a significant share of bandwidth it could use to create the foundation of a new wireless network. Unicom received 6 megahertz (MHz) of radio frequency in the 900 MHz

Table 3.1. Unicom's shareholders and their ownership percentage

Shareholder	Ownership Percentage[a]
Ministry of Electronics Industry (through ChinaCom–Huatong)	7.5
Ministry of Electric Power	7.5
Ministry of Railways	7.5
China International Trust and Investment Corporation (CITIC)	6
China Everbright ITIC	6
China Resources Group	6
China Huaneng Group	6
China Merchants Holdings Ltd.	6
China National Chemicals Import and Export Corporation	6
China National Technology Import and Export Corp.	6
China Foreign Economy, Trade, Trust, and Investment Corp.	6
Beijing Catch Communications Group	6
Shanghai Scientific and Technological Investment Co.	6
Guangzhou United Telecommunications General Corp.	6
China Fujian Foreign Trade Center	6
Dalian Vastone Telecommunications and Cables Co.	6

[a] Numbers are rounded off—7.5% is actually 7.46%, and 6% is actually 5.97%; these figures would give a total ownership percentage of 100%.
Source: *China Daily*, 19 July 1994, p. 12.

spectrum band—this was 50 percent more than the 4 MHz the MPT had for its own cellular phone network.[18]

Unicom's goals were ambitious: by the year 2000, the company sought to hold 30 percent of the country's mobile communications market, 10 percent of fixed-line communication, and 10 percent of long-distance service. In doing so, the company would lay some 20,000 kilometers of optical cable.[19] At its founding conference, its board chairman Zhao Weichen pledged the company would invest a rather astonishing 100 billion yuan ($12 billion) over the following five years to upgrade the telecommunications systems it inherited mainly from the power and railway ministries, and integrate them into the MPT's public telecommunications network.[20]

Despite these goals, and the implication that a new company would add catalytic rivalry to the telecommunications mix, the competitive role of Unicom was not clear from the beginning. In fact, according the State Council's regulations, Unicom only had the mandate to:

- 'Upgrade and perfect the existing private telecommunications networks of the Ministry of Railways and Ministry of Power... [and] to provide local telephone services to the areas where the public telephone networks cannot reach or there exists a severe shortage of telephone service.'
- Operate radio telecommunications business, including mobile.
- Operate value-added, undertake engineering services, and conduct other services related to the main business.[21]

According to a narrow interpretation of the company's mandate, Unicom was not necessarily meant to compete with the MPT. It could be seen only as filling gaps in the network built by its rival ministry, and in providing services in areas where the MPT was either absent or not yet a major presence.

The MPT itself had been hostile from the beginning to the creation of a rival. MPT minister Yang Taifang had apparently slowed the approval of an MPT rival company in the early 1990s, but new minister Wu Jichuan was too junior to halt the process when he took office in 1993.[22] The MPT took no ownership stake in the new company, and began active steps to retard its growth.

One of its first acts was to restrict Unicom's ability to cooperate with foreign investors in building its network. The ministry promulgated the 'Interim Methods on Approval of Providing Competitive Telecom Services' in September 1993, and 'Interim Rules on Market Management of Competitive Telecom Services,' in November 1995. These rules were meant to ban foreign companies from providing telecommunications services in China, and to prohibit domestic companies, such as Unicom, from providing such services in cooperation with foreign companies.[23]

However, the central government took a complementary step at the same time to reduce the power of the MPT. Since the early days of the reform era, the central government had advocated a program of *zhengqi fenkai*, or separation of direct government management from state-owned enterprises. Managers were in theory to be guided by profit motives and market forces, rather than by political mandates. For the telecommunications sector, this policy was made explicit (though not immediately enforced) in the 1979 State Council Directive number 165.[24] In 1994 the Council took steps to actualize the directive as it carved out the operating enterprise part of the MPT into a nominally separate corporation called the 'Directorate General of Telecommunications', or DGT.

The DGT, renamed as the company 'China Telecom' in 1995, emerged as the corporation to operate all aspects of the public communications network, while the MPT was to function as the nation's telecommunications regulator. Figure 3.1 gives a timeline of the telecommunications organizational structure from the 1950s to 2007 (companies formed since 1999 are discussed at the end of this chapter).

In theory, the separation of the MPT offices in 1994 was meant to make the ministry a more impartial regulator of the communications sector. However, as a state-owned enterprise, China Telecom still essentially came under the authority of the MPT, as major personnel decisions and expansion plans needed ministry concurrence.

Unicom, also a state-owned operating corporation, was nominally under the bureaucratic supervision of the State Council, but the MEI had major control over the development of the company. Though the MPT had separated itself from the former monopoly China Telecom, it still saw defending its corporation from the Unicom and MEI challenge as part of keeping influence over the course of the nation's telecommunications development. MPT local officials also maintained a mindset of monopoly control, though in more reform-oriented cities such as Guangzhou and Shenzhen, municipal officials were willing to encourage the growth of the rival telecommunications company.[25]

Competition between the two companies began almost as soon as Unicom started offering mobile phone services in July 1995. By that time, in its four target cities of Beijing, Shanghai, Guangzhou, and Tianjin, Unicom had invested some 700 million yuan ($84.3 million).[26]

Competition in the 1990s and 2000s

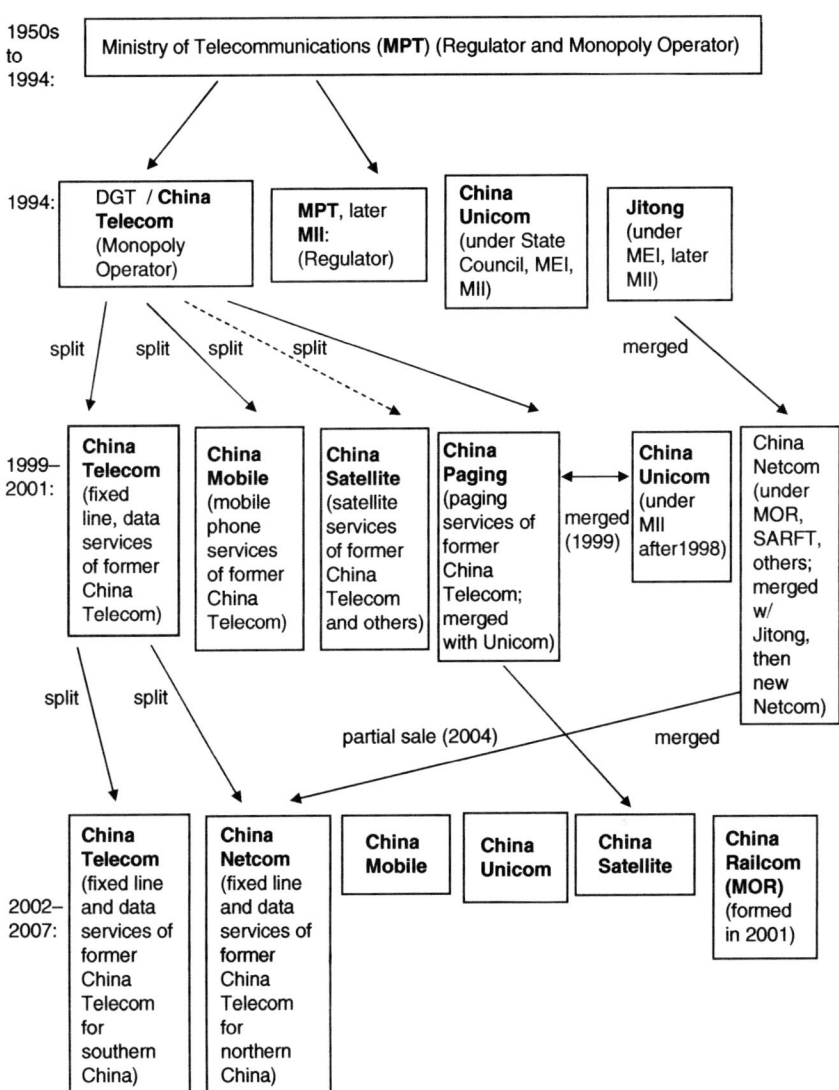

Fig. 3.1. Major players in telecommunications services, 1949–2007

The rivalry soon led to price wars between the established carrier and the fledgling company. In Guangzhou and Beijing, for example, following the debut of the new service option, the package price of installation, handset, and other charges fell from about 21,000 yuan (about $2,500) in early 1995 to some 10,000 yuan (about $1,200) in mid-1996. Shanghai saw similar price cuts as the two companies fought for customers.[27] Though the MPT was orchestrating efforts to try to eliminate its rival from China's large urban markets, the net effect was to benefit the consumer.

Following the completion of mobile networks in the first four cities in 1995, Unicom added an additional 13 cities in 1996, containing some 100,000 new customers. Networks for an additional twenty cities were also completed that year, and customer targets were surpassed in the cities of Shenzhen, Chongqing, Tianjin, Wuhan, Yantai, and Daqing.[28] The rapid expansion of the network was also a boon to Chinese customers around much of the nation, as it was no longer just those citizens in a handful of elite cities who had access to mobile phone choices.

The MPT quickly responded to Unicom's progress. It began by charging unreasonable rates for Unicom to use the national fixed-line and mobile phone system under its control, and by making it difficult for the company and its customers to connect calls made to those using the public network.[29] Even when connections could be made, Unicom had to pay high fees for using parts of the national phone network. For example, in Shanghai in 1997, Unicom gave 100 percent of its revenue from international calls and 92 percent of revenue from domestic long distance calls to the local city branch of the MPT. It was, however, allowed to keep about 92 percent of revenue generated by local calls to the MPT network.[30]

This tactic was a key weapon in slowing Unicom's growth, as Table 3.2 shows. Customers were disappointed that their new mobile handsets and phone subscriptions, while perhaps less expensive than those offered by the MPT's mobile service, often provided limited opportunities for reaching those they sought to call.

The telecommunications ministry was also determined to protect its monopoly on fixed-line telephone communication. Unicom, in

Table 3.2. Mobile phone customer growth and market share, 1994–1997, for Unicom and China Telecom, at year end

Year	Unicom	Unicom (%)	China Telecom	China Telecom / China Mobile (%)
1994	—	—	1.6 million	100
1995	10,000	0.3	3.6 million	c.100
1996	c.100,000	1	6.4 million	c.99
1997	339,000	3	9.7 million	97

Sources: Xinhua News Agency, 12 Jan. 1995 (for 1994); Xinhua News Agency, 25 Dec. 1995 (for Unicom, 1995); Xinhua News Agency, 18 Jan. 1996 (for China Telecom, 1995); South China Morning Post, 3 Oct. 1996, p. 8 (for Unicom, 1996): Xinhua News Agency, 17 Dec. 1996 (for China Telecom, 1996); Financial Times, 25 May 2000, p. 38 (for Unicom, 1997); Xinhua News Agency, 5 Feb. 1998 (for China Telecom, 1997).

cooperation with the US phone company Sprint, spent some 500 million yuan (about $60 million) to build a fixed-line network in the major city of Tianjin, and completed the network in July 1997. But local telecom officials said the 100,000-line network had technical problems, and refused to connect it to the MPT's China Telecom backbone. Unicom lost as much as 200,000 yuan (about $24,000) per day because of this.[31]

The MPT further battled to prevent Unicom from offering international phone connections. In mid-1997, the State Council ruled that the MPT could keep its monopoly on international service, as the ministry needed revenues to continue expanding its public network. The ministry argued that a price war between China Telecom and Unicom would whittle away at these funds.[32]

The MPT also appealed to the central government to criticize Unicom. The ministry pointed out Unicom's limited scope set by the State Council, and argued that Unicom should be using the networks provided by the Rail and Power ministries, and should not be wasting resources developing a network to compete with China Telecom.[33]

The cost competition and operating barriers hurt Unicom's balance sheet. For 1995, the company's operating costs were 55 million yuan, but income was only 11 million yuan, leading to a loss of about 44 million yuan (about $5.3 million). The following year, the company's Guangdong and Shanghai projects were profitable, and operations in Beijing broke even. The company's losses narrowed in 1996, and one source reported it made a profit of some 6.8 million yuan (about $800,000).[34]

How could the new company survive in the face of such fierce competition? One of Unicom's early tactics for battling against the MPT involved a creative scheme for gaining foreign funding and technical assistance. The company hoped this would give it an important edge to gain customers in the previously monopolized market.

Foreign Investment in Unicom

Early funding for Unicom came primarily from its government shareholder backers, as well as from loans from the People's Bank of China.[35] However, in the mid-1990s, the Chinese economy suffered from a bout of inflation, and the government tightened its spending. The State Planning Commission restricted Unicom's access to long-term capital, and required its approval if the telecommunications company wanted to spend more than 60 million yuan (about $7.2 million).[36] In 1995, then, Unicom began thinking of attracting foreign funds as an alternative capital source.

As previously noted, the Chinese government and MPT were loath to allow foreign telecommunications corporations rights to operate systems within China, and specific rules (but not laws) prohibited such cooperation. Unicom, however, decided to pursue a gray area of the policy, by setting up arm's-length joint ventures with foreign companies to both absorb foreign funding and acquire management expertise.

The investment scheme came to be known as 'Chinese–Chinese–Foreign', or CCF, agreements (in Chinese, *'Zhong-Zhong-Wai'*). Under CCF, Unicom represented the first 'C'. The second 'C' would in most cases be a satellite company from one of Unicom's shareholding entities. The satellite companies were distributed along a regional basis. For example, in Shanghai, shareholder China International Trust and Investment Corporation's (CITIC) Shanghai Trust and Investment Company represented the second 'C'; in Shanxi province, it was the MEI's subsidiary ChinaCom.

The foreign company ('F') would form a joint venture with the intermediary company to invest funds and cooperate in building the local or regional network; the contract could run for as long as twenty-five years. The foreign company could hold a majority stake in the venture, and was typically a 70–80 percent shareholder.[37] (For example, foreign investors Ameritech and Bell Canada both held 80 percent ownership in their joint ventures in Shanxi and Shandong provinces).[38] The joint venture would then sign a contract of as long as fifteen years to provide services to Unicom.[39] In doing so, the foreign company could establish an indirect link with Unicom, while avoiding the prohibition on direct investment in the Chinese telecommunications carrier.

At the time the CCF's developed, there was no telecommunications law on the national books to regulate such dealings. One Unicom official noted in 1995 that the company 'cannot wait for a law', and that it depended on the State Council's support in moving ahead with their project. The official indicated that State Council support was more important than a law.[40]

Table 3.3 indicates the regional distribution of Unicom's shareholders. Table 3.4 gives a partial list of CCF projects. In all, Unicom created more than 40 CCF ventures over the years 1995 to 1998 when, as we will see, the scheme was cancelled by the MPT's successor ministry.

Because of MPT restrictions, foreign firms were not allowed to have a direct role in management or oversight of the CCF companies. However, according to interviews with foreign representatives of some of the foreign partners,

Table 3.3. Partial list of Unicom's regional shareholders

Unicom Shareholder	Province / City
China National Tech. Import/Export Corporation (CNTIEC)	/ Beijing
China Fujian Foreign Trade Center	Fujian
CITIC	/ Tianjin
CITIC	Liaoning
CITIC	Shandong
CITIC	Hubei
Shanghai Science and Tech. Investment Corporation (SSTIC)	/ Shanghai
MEI (ChinaCom)	Shanxi
MEI (ChinaCom)	Jilin
MEI (ChinaCom)	Hunan
MEI (ChinaCom)	Jiangxi
China Huaneng Group	Sichuan
China Resources Group	Guangxi
Ministries of Railways, Electric Power, and MEI	Yunnan
Guangzhou South China Telecom Investment (GSCTI)	Guangdong
China Merchants Holding Co.	/ Shenzhen
Beijing Communications Group[a]	Hebei

[a] Beijing Catch owned 85 percent of the regional Hebei branch, Hebei United Telecommunications Equipment.
Sources: Paul Triolo, 'China's United Telecommunications Corporation: New Leader, New Organization, Old Problems', Report for American Embassy in Beijing, May 1996,' 3–4; *South China Morning Post*, 20 Aug. 1999, Business Post, p. 4; interview with Ameritech official, June 1996.

Unicom would often turn to the outside companies for advice on equipment procurement, network design, and related management matters.[41]

In addition to receiving technical assistance from foreign companies, Unicom's main goal was to absorb foreign capital through the CCF scheme. In this, the company was quite successful, as more than twenty foreign companies contributed as much as $1.4 billion in the more than forty joint ventures up to 1998.[42] Over the three years from 1994 to 1997, some 70 percent of Unicom's capital came from overseas.[43] Unicom had raised a useful amount of funds, but had fallen far short of its 1994 goal to invest as much as $12.5 billion to upgrade its network.

What did the foreign companies stand to gain from the investments? Chinese regulations barred direct revenue sharing—this was the primary reason the cooperation could not be a one-to-one joint venture. So foreign companies were to receive a portion of operating revenues in return for their seemingly intangible contributions to network development.

China Telecom itself was not allowed to take part in CCF projects, as the MPT remained suspicious of Unicom's foreign fund-raising activities. This meant that any foreign telecommunications operating company had only Unicom as a target, and many companies disregarded the unclear remuneration methods and lack of explicit MPT approval to cooperate with Unicom.

Competition in the 1990s and 2000s

Table 3.4. Examples of Unicom 'CCF' joint ventures with foreign firms

Foreign Firm	Chinese Partner[a]	Target Region
Ameritech (USA)	ChinaCom	Shanxi
Asian American Telecom. Corp. (USA)	Huaneng	Sichuan
Bell Canada (Canada)	CITIC	Shandong
Bell Canada (Canada), AIG (USA)	Hehua Elec. Information Group[b]	Shandong
CCT Telecom (Hong Kong)	SSTIC	Shanghai
CCT Telecom (Hong Kong)	SSTIC[c]	Shanxi
Daewoo (South Korea)	n/a	Zhejiang
Deutsche Telecom (Germany)	CITIC	Tianjin
France Telecom (France)	GSCTI	Guangzhou/Foshan
MasterCall (Thailand)	CNTIEC	Beijing
McCaw International (Nextel) (USA)	SSTIC	Shanghai
NTTi and Itochu (Japan)	Hebei United Telecom. Equipment	Hebei
Singapore Telecom International (Singapore)	SSTIC	Shanghai
Singapore Telecom International (Singapore)	Zhongshan Group[b]	Suzhou
Stet (Italy)	ChinaCom	Jilin
Telesystem International (Canada)	ChinaCom	Hunan

[a] In some cases, the local partner was itself a subsidiary of the Chinese parent Unicom shareholder. For Bell Canada, for example, the local partner was CITIC's Yantai Industry and Trade Corporation.
[b] The Hehua Electronic Information Group manufactured electronic equipment, and the Zhongshan Group was a manufacturer of electronics and telecommunications equipment; both companies' links to an original Unicom shareholder are not clear.
[c] The CCT Telecom venture in Shanxi was made via its initial investment in its venture in Shanghai. The company apparently took over the stake in Ameritech's investment, which ended in 1997.

Sources: *South China Morning Post*, 16 June 1995, Business section, p. 10; *Asia Pacific Telecoms Analyst*, 9 Oct. 1995, p. 16; *Straits Times* (Singapore), 22 Mar. 1996, Money section, p. 44; *Wall Street Journal*, 3 June 1996, p. B4; Horizon House Publications, Inc., 'China: Towards the World's Largest Market', *Telecommunications International Edition* 30 (1996): S1–S34; Bell Canada website report, **http://www.bci.ca/en/html/communications/morepr/press_releases/1997/97.11.28.html**; China Online, 2 Sept. 1999; *South China Morning Post*, 20 Aug. 1999, Business Post, p. 4; *South China Morning Post*, 11 January 2000, Business Post, p. 1; *South China Morning Post*, 2 February 2000, Business Post, p. 3; *South China Morning Post*, 31 May 2000, Business Post, p. 3.

The case of Ameritech illustrates the approach foreign companies took to Unicom in the mid-1990s.[44] The American carrier (which merged in 2000 with SBC Communications) first looked to partner with the MPT's regional phone companies, but was rebuffed. It then turned to Unicom. Though the MPT rival had initially signed many rather vague MOUs with foreign companies, Ameritech was eager to sign a contract. The American company opened a representative office in Beijing in April 1995, and began talks with both Unicom and one of its branch partners, ChinaCom (also known as Huatong).

ChinaCom's first offer was for cooperation on developing a mobile network in just one city, Taiyuan, in its region of Shanxi province. But Ameritech requested, and was given, rights to the whole province, though the early targets were two large cities, Taiyuan and Datang. Ameritech and ChinaCom signed a fifteen-year cooperative joint venture contract on 10 August 1995, with an investment of $20 million and registered capital of $8 million. The goal was to construct a 10,000 subscriber digital cellular system.[45] Ameritech took a controlling interest of 80 percent of the company, though it was mandated to reduce its stake to 49 percent once it had recovered its investment. The company expected operations in Shanxi would begin sometime between 1999 and 2001.

Ameritech's plans were cut short (perhaps fortuitously, as we will see later) in July 1997, as it became the first foreign company to leave China's telecommunications market. It is not clear exactly why the company left. Apparently, directors in the US felt the prospects for achieving a return on the investment were not bright in the short term. According to reports, the company's contribution to the venture had not included any cash, so Ameritech emerged relatively unscathed from its China project.[46]

It is noteworthy that, as Table 3.4 indicates, there was relatively little overlap between foreign investments in the same city or region. Unicom seemed uninterested in pitting foreign companies against each other in major cities such as Beijing or Guangzhou, although both Singapore Telecom and McCaw International had investments in Shanghai. One problem, of course, was that the foreign companies would have to partner with the same regional branch of Unicom, and it seems that Unicom's rules generally permitted only one foreign company per region—perhaps the 'gray area' nature of the practice inspired some caution. However, Unicom was thereby unable to benefit from potential competition between foreign investors for one local market, even if the competition was done at arm's length.

Though Unicom was able to secure many interested partners that lasted beyond 1997, as Table 3.4 indicates, the level of foreign investor satisfaction with the company was quick to fall. One core problem was that the MPT was slow to allow interconnection of the Unicom network with that of the national phone system—this tactic imitated the American company AT&T's strategy of isolating potential rivals in the early twentieth century (see Chapter 2). The president of Telesystem International's venture in Hunan, for example, complained it took eight months to connect his company to the public network, resulting in a loss to Unicom of 19 million yuan (about $2.3 million).[47]

Other investors chafed at the inability to participate in operations, pointing out that Unicom's billing systems, equipment maintenance, and other service functions were sorely lacking in the four mobile phone sites (Beijing, Shanghai, Guangzhou, and Tianjin) that went into operation in July 1995.[48] One foreign executive complained that no BOT (Build–Operate–Transfer) schemes existed in the industrial sector; he facetiously called the arm's-length Unicom investment plan 'BTT,' for 'Build, Transfer, Thank you very much.'[49]

Central Government Regulation to Maintain Competition

Central government actions in the mid-1990s to support Unicom showed there was concern about the tactics the MPT was taking to slow the junior carrier's progress. In late 1995, a 'leading group' of top government officials was formed to examine high technology policy making. It was led by vice premier Zou Jiahua and orchestrated by MEI chief Hu Qili, and seemed to indicate the development of the industry might be headed to levels somewhat beyond that of the battling ministries.

In early 1997, China's State Economic and Trade, State Planning, and State System Restructuring Commissions (all superior to the MPT in the bureaucratic pecking order) composed a report for then-vice premiers Zou Jiahua and Wu Bangguo.[50] The report provided a thorough review of the development to date of the rivalry between Unicom and the MPT. Among other assertions, it solidified the idea that competition within the telecommunications sector had become a fact of life:

Unicom and China Telecom are both telecommunications service industry state-owned enterprises, so under the industrial management of the Ministry of Posts and Telecommunications, they are responsible for managing their own networks, and within the socialist market economy *they should both compete and cooperate* in order to positively advance the nation's competitive development process.[51]

The document further addressed basic conflicts that had grown between the two telecommunications carriers. The MPT asserted Unicom's creating a nationwide rival network violated the conditions set out in the State Council Directive 178, and implied the company was duplicating the ministry's efforts. The MPT also asserted Unicom should restrict itself to the Rail and Power ministries' networks and to areas not covered by the MPT, as approved in its charter, and also stay away from international long distance and fixed-line local calls.[52]

The MPT also argued the virtues of continuing a monopoly for most of the telecommunications network. It cited, for example, the national security benefits of having only one company to regulate international calls; in this way, the government could more easily monitor communications of Chinese citizens with people outside the country. A monopoly would also help the country by preventing foreign companies from using the competition as leverage against the nation's telecommunications system, and would allow China to present a united front on issues such as bargaining with foreign governments on payment for international call completion.[53]

Unicom responded that it needed access to the foreign currency generated by completing international calls in order to pay foreign operating companies that were financing and beginning to cooperate with the new company (discussed below and in Chapter 6). It also pointed out that other countries had competing telecommunications companies, so that a monopoly was not necessarily in the best interests of the country. Furthermore, it claimed it could use new technology to build an efficient network, one that would provide good service and satisfy the needs of a growing market. Unicom also argued that it was charging a lower rate for its calls than was China Telecom, and that this was a beneficial service to the country's users.[54]

The 1997 report did not clearly support either Unicom or the MPT, but did encourage codifying Unicom's ability to compete with the MPT in a variety of telecommunications areas. For example, it called for the MPT to be an impartial regulator of the sector, and allow the systems built by the Rail and Power ministries fair access to the national network. Unicom should also have at least a trial access to providing local fixed-line telephone calls, and should be allowed to charge up to 10 percent less than China Telecom for its telephone calls. As noted above, the powerful economic commissions also reasserted their larger support for competition within the sector.[55]

The government did move to modify the way foreign investors could take part in Unicom's development. The report had criticized the CCF framework for allowing foreign companies to take a 70–80 percent stake in their part of the cooperative effort—if there were profits, Unicom would have to make a big pay out to the foreign participant.[56] A pilot project announced in May 1997 was designed to allow a foreign company a direct joint venture with Unicom, as long as the investment was less than 50 percent.[57]

Unicom's Internal Problems in the Mid-1990s

Despite the obvious competitive dynamics inspired by bureaucratic rivalry, some of the very same factors that allowed the creation of Unicom soon came to work against the corporation. Incompetent early leadership and competition among Unicom's shareholders made it more difficult for the company to coordinate its early growth.

Unicom's first chairman was selected by the State Economic and Trade Commission (SETC), and, apparently to ensure impartiality, could not come from the MEI, MOP, MOR, or any of the other major shareholders.[58] Zhao Weichen, a former vice governor of Guangxi province, who had also worked under then-vice premier Zhu Rongji at the SETC, took the company's top position.

Zhao seemed a curious choice for the job. He had studied at Beijing's prestigious engineering university, Tsinghua, and worked for many years in the machine-building bureaucracy. But he had little real connection to the telecommunications sector, and certainly a far weaker background than that held by his chief rival at the MPT, Wu Jichuan. It is likely his SETC connection made him the prime candidate for the job.

As noted in the earlier discussion of foreign investment, the MPT's prohibition on forming joint ventures with any foreign-related entity meant Unicom received lavish attention from Western operators seeking entrée to the China market. Zhao exploited the foreign interest, and was plied with constant offers of elaborate banquets sponsored by foreign suitors and all-expenses-paid travel around the globe. He apparently had little time left to devote to the company's management or development, even with former American secretary of state Henry Kissinger helping as a senior consultant.[59]

One foreign investor criticized Zhao's laxity, asserting 'he would hand out MOUs over lunch', and claimed that Zhao had signed more MOUs for Guangzhou than there were licenses.[60] Indeed, in 1995 Unicom signed some sixty-six MOUs, with twenty turned into actual contracts and sixteen expiring. The number dropped to only twenty-one in 1996, as foreign interest and trust in the 'CCF' model began to wane.[61]

In March 1996, Zhao Weichen was replaced as head of Unicom, and another SETC official, Yang Changji, took over briefly as company chair. But later that year the Ministry of Electronics acquired the duty of naming a new leader for the company, and MEI vice minister Liu Jianfeng took Yang's job.

The real voice for the company on Zhao's departure, however, was Li Huifen, a former radio equipment company manager and vice mayor of Tianjin, and alternative member of the Communist Party's Central Committee. She was given the post of company president in addition to her State Council-appointed job as general manager, and she controlled the company for the next two years.[62]

Li, like Zhao, a graduate of Tsinghua University, with an engineering background, was a strong believer in Unicom's mission to rival the MPT monopoly. She particularly challenged the ministry's sloth in granting connection rights to Unicom users, saying in a public speech that 'Unicom has to overcome great difficulties to advance one step', and the company had to fight to get its first four cities connected. She complained that by mid-1996 the company had made 'little progress' in getting its next batch of twenty targeted cities access to the public network.[63]

During Li's early tenure, the company moved to simplify the way foreign investors could take part in Unicom ventures; as noted above, the program of arm's-length joint ventures was modified to allow direct foreign minority stakes in a project.[64] She also talked about listing the company on Hong Kong's stock exchange, and pushed to open the company's fixed-line network in her former home city of Tianjin.[65]

Despite the revived leadership under Li, rivalry among the Unicom partners also weakened the company. By 1997, ChinaCom (or Huatong), the MEI affiliate, had taken over the ownership share of one of the original thirteen smaller partners, the China National Chemicals Import and Export Corporation, giving it the largest ownership share of about 13.4 percent. The MEI unit reportedly demanded special privileges from Unicom because of this ownership stake.[66]

Unicom's vice presidents were appointed by the company's other two main ministerial investors, the Power ministry and the Railway ministry. The two vice presidents apparently competed with each other in 1997 to get foreign contracts, thus complicating Unicom's ability to speak with one voice. Li Huifen was reportedly concerned with her own political future beyond Unicom, and declined to mediate among the competing branches of the company.[67] Moreover, even after her replacement of Zhao Weichen, the company was known to promise the same project to multiple potential investors.[68]

Despite efforts to reform Unicom and solve its problems under its original organization structure, the Fifteenth Communist Party Congress of September 1997 brought unfavorable personnel news to the company. Unicom's highest-ranking political proponents, MEI minister Hu Qili and Politburo member Zou Jiahua, both lost their important Communist Party Central Committee seats. At the same time, MPT minister Wu Jichuan was promoted to be a full member of the Central Committee. These changes meant two key Unicom supporters were now neutralized, and a rival saw his power enhanced.

Given the multiplicity of problems and challenges Unicom faced by the end of the year, the company stood little chance of achieving its early growth targets, and it was rapidly falling behind a revitalized MPT communications machine. Unicom had constructed an unworkable management hierarchy inspiring constant shareholder tensions, had alienated some important foreign investors, and was under constant pressure from manipulative MPT practices.

Foreign investors, though still skeptical of the new company, seemed willing to wait for an improved investment climate, one that stood to change with the proclamation of a perpetually delayed telecommunications law. Moreover, the collapse of Unicom would mean the MPT would again be a monopoly, and the attempt to introduce competition would be a visible failure.

The following year, however, was to see both a profound change and a chance for resurrection for the rival telecommunications company.

Unicom's Transformation in 1998

A major turning point for Unicom came in March 1998, when Zhu Rongji took China's prime minister position with an intent to consolidate the country's bureaucracy by merging and closing ministries. At the same session of the National People's Congress that promoted Zhu, the MPT and MEI were amalgamated to form a new Ministry of Information Industry (MII).[69] At the same time, the Ministry of Radio, Film, and Television was renamed the State Administration for Radio, Film, and Television (SARFT), which, as Chapter 4 notes, came to be a potentially new rival to the MII as it strove to offer data services through its cable television network.

The choice of new MII leadership would set the pace for future development of the whole telecommunications industry. The post went to the MPT's Wu Jichuan, who apparently saw his only challenge from Unicom chief and MEI vice minister Liu Jianfeng, who became an MII vice minister. The MPT's Yang Xianzu and Zhou Deqiang (Wu's ultimate successor as MII minister) also became vice ministers, as did the MEI's Lu Xinkui and Qu Weizhi.

Within months, much of the MEI and Unicom's former leadership was rooted out of the MII. Liu Jianfeng served only two months in the new MII, then went on to head China's civil aviation administration until 2002.[70]

Competition in the 1990s and 2000s

MEI minister Hu Qili was sidelined as well, becoming a vice-chairman of the largely ceremonial Chinese People's Political Consultative Conference, though he did take a post at the Shanghai Huahong Electronics Corporation. Li Huifen gradually lost her remaining leadership authority in Unicom, though it was not until February 1999 that Yang Xianzu officially took the chairman's role at Unicom, effectively consolidating MPT control over the company. Wang Jianzhou, director of the MII's planning department, became Unicom's executive vice president.[71]

The bureaucratic reshuffling had a startlingly quick affect on Unicom's fortunes. As part of the MEI, Unicom was a rival; now, under the MII umbrella, it could find room to grow. By June 1998, for example, Unicom's fixed-line network in Tianjin had received permission to connect to the national network.[72] In late 1998, Unicom received a license to conduct its mobile phone business nationwide, thereby putting it on the same footing as its rival China Telecom.[73]

At the same time, the company's new masters came seriously to question the CCF scheme and the deals that had been made under the leadership of Zhao Weichen and other Unicom executives. Wang Jianzhou, just before his own move from the MII to Unicom, said the company had been investigated in 1998 for 'unprincipled and wrong foreign fund-raising practices'. New MII chief Wu Jichuan commented in October that 'China Unicom has started to use the Chinese, Chinese, Foreign method... but in this method we have found many irregularities'.[74]

There were many reasons the CCF deals were being called into question. First, the MPT and new MII leadership had all along opposed a foreign role in operating telecommunications systems and taking revenue when they became profitable, and now Unicom had come under their control. Second, the CCF arrangements themselves were by their nature not entirely transparent, causing the ministry to question the flow of funds from Unicom to the foreign investors.

Perhaps the most important reason for re-examining the CCF contracts was a drive in 1998 and 1999 to list Unicom on an overseas stock exchange. Rival China Telecom had listed part of its mobile phone assets in Hong Kong in 1997, and the government saw readily available foreign funds without allowing foreigners actually to own a controlling stake. In order for Unicom to conduct its own overseas listing, however, the company would have to clarify its obligations to foreign investors. The fastest and most convenient way to resolve this, as well as other issues related to the CCF pacts, was simply to declare them void. Foreign companies were finally notified by September 1998 that their agreements were to be banned, though the disposition of the contracts was not initially clear.[75]

Soon, the foreign companies expressed their dismay to the media and their embassies, and some twenty companies tried to establish leverage by forming a 'Unicom Investor Group'. Members included Singapore Telecom, Sprint, and France Telecom.[76] According to one account, the MII tried to 'divide and conquer' these foreign companies by telling them that companies that settled early would get better compensation. Some were warned that those who refused to end their contracts would be prohibited from future participation in the China market.[77]

Eventually, the foreign investors saw they had little choice. In a 30 August 1999 directive, the MII informed Unicom that 'the CCF project contracts... are in violation of current State policy and regulations, and must be corrected'. The directive instructed Unicom that 'in principle, the projects in violation of State regulations shall be liquidated prior to mid September, or at latest, by the end of September, and the liquidation results shall be reported to the Ministry of Information Industry'.[78] Unicom in turn told its former CCF partners that on 1 October 1999 it would stop paying any revenue.[79]

One of the first contracts to be dissolved was the Hebei venture created by Japan's NTT and Itochu companies, which announced termination in September 1999. The Japanese investors apparently received the original cost of their investment plus a 7 percent return—this was apparently the offer for companies that sold out in the early stages.[80]

CCT Holdings of Hong Kong was one of the few winners in their settlement. In February 2000, the company received from 800 million to one billion yuan ($96–120 million) on its initial 600 million yuan ($72 million) investment, along with options to buy shares of Unicom on its planned public listing.[81] Metromedia (which had acquired Asian American Telecommunications Corp.) also did well, receiving in late 1999 more than 800 million yuan ($96 million) for four CCF joint ventures; this was apparently 28 percent more than the initial investment.[82] In February 2000, Bell Canada reached an undisclosed settlement.[83] However, Telesystem International sold their stake in their Hunan venture in April 2000, for about 209 million yuan ($25.2 million), apparently recovering only 95 percent of the purchase price.[84]

The settlements helped clear the way for Unicom's initial public offering (IPO). The listing was originally scheduled for late 1999, and the Chinese government approved it to raise $1 billion. The MII had insisted the CCF deals be resolved before the IPO could proceed.[85] With resolution of the compensation issue, Unicom applied in April 2000 for a dual listing on the Hong Kong and New York stock exchanges.

Continued Competition under the MII

China Unicom followed a pattern similar to that of China Telecom in structuring its public offering, creating a new 'China Unicom' within the parent company that used the formal English name of China United Telecommunications Corporation to distinguish itself.

Unicom's IPO in June 2000 was a spectacular success. The company raised $4.91 billion by selling only 122.95 million shares, or 5 percent of its total. The amount raised was a record, surpassing China Telecom's mobile phone listing in 1997 that had garnered $4.5 billion.[86]

The transfusion of funds seemed to revitalize Unicom. As Table 3.5 indicates, subscriber numbers grew rapidly. Of course, other factors contributed to the exploding demand for mobile phones in the early 2000s. Technological advances had led to a rapid fall in the cost of equipment, rising levels of income in China had made the service more affordable, and

Table 3.5. Mobile phone subscriber growth and market share, 1997–2007, selected years, for Unicom and China Telecom / China Mobile,[a] at year end (except as noted)

Year	Unicom	Unicom %	China Telecom / China Mobile	China Telecom / China Mobile %
1997	339,000	3	9.7 million	97
1998	1.4 million	6	23.9 million	94
1999	4.2 million	9	40 million	91
2000	12.8 million	15	72 million	85
2002[b]	40 million	26	114 million	74
2003[c]	79 million	36	139 million	64
2005[d]	123 million	35	231 million	65
2007[e]	158 million	31	356 million	69

[a] In May 2000, China Telecom's mobile services were transferred to China Mobile.
[b] As of Oct. 2002.
[c] As of Nov. 2003.
[d] As of Aug. 2005.
[e] As of Oct. 2007.

Sources: Financial Times, 25 May 2000, p. 38 (for Unicom, 1997 and 1999); Xinhua News Agency, 5 Feb. 1998 (for China Telecom, 1997); Telecommunications Development Asia–Pacific website, http://www.tdap.co.uk/uk/archive/interviews/inter(unicom_0012).html (for Unicom 1998); Financial Times, 10 Nov. 1999, p. 1 (for China Telecom, 1998); China Daily, 30 Jan. 2000, p. 5 (for China Telecom, 1999); China Daily, 22 June 2001, p. 5 (for Unicom, 2000); AsiaInfo Daily News, 27 June 2001 report (for China Mobile, 2000); Financial Times Information, Global News Wire, 21 Nov. 2002 (for 2002); Financial Times Information, Global News Wire, 19 Dec. 2003 (for China Mobile, 2003), and 23 Dec. 2003 (for Unicom, 2003); Comtex News Network, 23 Sept. 2005 report (for 2005); Xinhua Economic News Service, 20 Nov. 2007 report (for 2007).

Competition in the 1990s and 2000s

growth of businesses and consumer mobility in general fueled citizen need and demand for mobile phones.

As Table 3.5 also indicates, the incorporation of Unicom under the wing of the MII made the ministry less reluctant to cede market share from its own mobile phone company, and by 2002 Unicom had more than a quarter of the national mobile phone market. The rest of the market share was still held by the only other company allowed to operate mobile phone systems, China Telecom, which split off its mobile phone operations as a new company, China Mobile, in 2000 (China Mobile is discussed in more detail below).

Despite the nominal union of the companies under the MII, competition between China's major mobile phone carriers continued within months of the ministerial merger. In May 1999, for example, Unicom's Wuhan city branch introduced a mobile phone service that required users to pay only a one-time 150 yuan ($18) connection fee to use the network, and have local calls charged at 0.2 yuan (about 2.4 cents) per minute, which was half of the normal rate. Callers were not, however, able to use the phones outside the city.

Wuhan's local branch was flooded with customers requesting the service, which the company had apparently been using in other cities such as Shanghai, Changchun, and Chongqing. China Telecom protested that the rates violated national pricing laws. Wuhan's municipal Price Bureau ordered Unicom to raise the connection fee to 500 yuan and the per-minute charge to 0.4 yuan, and Unicom complied, though under protest.[87] However, the similar services in the other major cities noted above were apparently unaffected.

In mid-1999, the government also opened voice over Internet protocol (VOIP) pre-paid card services, which allowed use of the country's data network for voice telephone calls, to China Telecom, China Unicom, and the Internet network company Jitong. At the beginning of the program, the companies were not to compete on price.[88] But in January 2001, new competitor China Netcom (discussed in more detail below and in Chapter 4) cut many of its IP telephone card prices in half, causing Unicom, China Telecom, and Jitong to match the new rates or face the loss of customers.[89] In July of that year, the State Development Planning Commission ratified the competition, as it told the MII the forced price control should end.[90] By 2002, all of the companies were offering further VOIP prepaid card discounts of as much as 30–40 percent in some of China's major cities.[91]

Competition in the 1990s and 2000s

Unicom also made a short-lived attempt to add paging services to its portfolio. In mid-1999, the company absorbed Guoxin Paging, itself split off from China Telecom in 1998. Guoxin had combined ten paging companies formerly directed by the MPT.[92] However, the paging business turned out to be a losing venture for Unicom, and in early 2004 it sold part of the company to China Satellite Communications (aka Satcom) (discussed below).[93]

Finally, Unicom chose to be China's first company to try to introduce a new technical standard of mobile phone service, known as CDMA (code division multiple access). CDMA was said to be technologically superior to the GSM (Global System for Mobile) system both China Telecom and Unicom had been using. It promised more efficient use of network capacity and, for customers, potential for rapid transfer of mobile text and data, as well as limited access to Internet web services.

By early 1998, Unicom had begun building CDMA networks in Tianjin, Shanghai, and Guangzhou, while China Telecom kept to its GSM service.[94] But the CDMA service languished for several years. By January 2002, the company had spent some 24 billion yuan ($2.9 billion) on building its CDMA facilities.[95] But the new technology had attracted fewer than 100 customers in the major market of Beijing. At prices of some 4,500 yuan ($540) for a handset, the cost of the new system was far higher than that of GSM.[96]

By mid-2004, Unicom had invested some 50.4 billion yuan (about $6.1 billion) in three stages of building the new network. The company won subscribers by subsidizing the cost of the handset up to 456 yuan ($55) per phone, and new low-end models cost as little as 700 yuan ($84).[97] Customers also began to see the advantage of CDMA for using wireless Internet services such as playing games and sending photos. (Internet use is discussed in more detail in Chapter 4.) By the end of 2005, Unicom had attracted about 33 million CDMA customers nationwide, though it had ended its program of subsidizing handset costs. Unicom's GSM service was still its core, and claimed some 95 million users.[98] In mid-2007, of Unicom's 150 million subscribers, only 39 million had signed up for the CDMA service.[99]

Meanwhile, as the following section will discuss, China Mobile had delayed moving into a new generation of mobile phone standard, apparently waiting for the 'third generation' (or '3G') standards, which promised rapid access to data and Internet services on the mobile handset. But while CDMA's long-term potential contribution to Unicom was uncertain, it was already clear that Unicom and its major competitor, China Mobile, had taken different paths even as they worked under the same MII.

New Entrants to the Telecommunications Field

China Mobile: facing a new competitor in the 2000s

China Mobile was officially separated from China Telecom in May 2000. The new company took all of the former China Telecom's wireless services. At the company's founding, MII leader Wu declared that the Chinese government wanted 'orderly and cooperative competition'. According to the official Xinhua news agency, 'all operators have expressed their willingness to avoid duplicated construction and vicious competition'.[100]

The new personnel leadership line-up for both China Telecom and China Mobile seemed to guarantee continued MII control. The former president of China Telecom, Zhang Ligui, simply moved to China Mobile as both president and party chief. The remaining, fixed-line China Telecom took as president and party leader Zhou Deqiang, a vice minister at the MII and Wu's successor as minister. Many other top leaders of both companies had affiliations with the ministry.[101]

China Mobile continued its drive to attract customers to its GSM service, and, as Table 3.5 indicates, its customer base grew rapidly in the new decade. In 2000, it began building a network to provide GPRS (general packet radio service) connections for it customers. GPRS, like CDMA, offered customers improved data services, with download speeds up to 115 kilobits per second (kbps), twice as fast as a home computer modem's 56 kbps.[102] As of mid-2004, the company had about 20 million GPRS customers.[103]

China Mobile also continued to compete with Unicom on price. In 2002, for example, one report indicated the company had quietly offered promotional packages to undercut Unicom. In Shandong, rates were less than half of what the MII would officially allow.[104]

Just as the two wireless carriers seemed to settle as duopoly competitors, a new technology came to present a challenge to both mobile phone companies. In January 1997, the State Radio Regulatory Committee approved frequency to be used in a new communications standard from Japan, called PHS (for 'personal handy system').[105] The technology allowed essentially fixed-line phone customers to roam within their local phone area, with handsets used like mobile phones, but charges basically the same as their fixed line. Though the quality of the connection was not up to that of traditional mobile or fixed-line phones, the calls were billed at local phone rates, and, in contrast to the regular mobile phone system, only the calling party paid.

By May 2000, there were about 600,000 China Telecom subscribers of what the Chinese came to call *xiaolingtong* (usually translated as 'Little Smart') in fifty cities, though large cities such as Beijing and Shanghai would have to wait a few years to offer the service.[106] Significantly, however, China Telecom did not have a license from the MII to run mobile phone operations, and the technology was therefore seen as an extension of fixed-line service.

At the end of 2000, the MII raised the per-minute charge to 0.2 yuan (2.4 cents) per minute, up from the same rate for three minutes. According to one official, the ministry made the move to discourage the spread of the technology, and, implicitly, to support the mainstream China Mobile services.[107] However, traditional mobile rates were still about twice those of *xiaolingtong*.

The rate move did little to curb enthusiasm for the quasi-mobile system, and, despite continued quality problems, it grew to 5 million subscribers at the end of 2001, 15 million in mid-2003, and more than 77 million (split between China Telecom and the new fixed-line carrier, China Netcom) as of May 2005.[108]

Despite the limited services and lower quality of *xiaolingtong*, China Mobile did see the service as a competitor. In 2002, after China Telecom launched a PHS network in the major city of Chongqing, China Mobile offered a competing service called 'City Smart' that cut the normal rate in half, and offered local calls for 0.2 yuan to 0.3 yuan per minute.[109]

Even as China Telecom had seemingly given up its mobile services, *xiaolingtong* gave it new life in the mobile phone arena. The company represented another key competitor to China Mobile, one that offered a choice of services and price to the Chinese consumer.

Railcom and Satcom: the minor competitors

At the beginning of the decade, the Chinese government saw the benefits of competition, and moved to allow even more entrants to the communications service field. But the cases of Railcom and Satcom indicate that companies that begin life with few comparative advantages could find little chance to grow in a market where monopoly inefficiency had already begun to disappear.

Unlike mobile operations, fixed-line phone service remained in the monopoly hands of China Telecom throughout the 1990s (see Figure 3.1). As noted above, however, some specialized fixed networks already existed outside of China Telecom's control, such as those owned

Competition in the 1990s and 2000s

by the Ministries of Power and of Railways, both investors in China Unicom.

The Ministry of Railways was not content with its small stake in China Unicom, and after the MII had taken control of the mobile phone challenger in 1998, the MOR soon made noises about setting up a new rival to China Telecom. In 1999, the ministry controlled a nationwide network of 120,000 kilometers of telecommunications lines, and claimed it had plans 'to establish a large telecommunications company that will be able to do exactly what China Telecom does'.[110] The ministry's 26,000 kilometers of optic fiber cable were second in the nation only to China's Telecom's 180,000 kilometers. However, with only 1.5 million lines, the ministry at the time had only about 1 percent of China Telecom's capacity.[111]

In 2000, the MOR held talks with China Unicom about merging their assets, but the MII apparently worked against the move that would have added the burden of a heavily over-staffed company to the mobile carrier's payroll.[112] The MOR employed some 65,000 people in jobs essentially guaranteed by old-line socialist state employment practices, while Unicom had only about 18,000.[113] Still, the central government favored creation of a rival fixed-line network, and Premier Zhu Rongji was a strong supporter for creating a competing company to China Telecom.[114]

Following the failed merger attempt, the MOR received a license to operate its own independent fixed-line communications network, and in early 2001 established China Railway Telecom, or China Railcom. In addition to overstaffing, Railcom also suffered a shortage of trained telecommunications specialists. The MII only allowed Railcom to offer a maximum 10 percent discount in costs to China Telecom service, and by the end of 2002, interconnection of Railcom's network with that of the former monopoly had made little progress.[115]

Railcom floundered in the early years of the decade, attracting few customers despite offering phone rates often lower than those of China Telecom. It was distracted by having to service both new private customers as well as its traditional railway-related users. It sought a merger partner, but Unicom continued to rebuff the company, and China Telecom and later Netcom saw little profitability in acquiring it. China Mobile, which in late 2003 was also rumored to be interested in the fixed-line company, also decided not to take it over.[116]

Finally, in January 2004, the MOR lost direct ownership of Railcom, and the company's assets were moved under the direct control of the newly formed State-Owned Assets Supervision and Administration Commission (SASAC).[117] In the middle of the year, the company was re-christened 'China Tietong'. The MOR was left to repay about $286 million in Tietong debt, and the ministry would also shoulder the telecom company's foreign debts to the World Bank and Swiss Bank. The new company held only about 3 percent of China's fixed-line market (and only about 1.5 percent of total phone subscriptions, including mobile).[118]

As of late 2005, Tietong had nearly 15 million fixed-line subscribers, and about 1.65 million customers for its broadband CRNET Internet service (discussed in more detail in Chapter 4).[119] The subscriber number remained small, still only about 2–3 percent, with some 2 percent of China's broadband subscribers.[120] Tietong remained frustrated by having to compete with the giant rivals China Telecom and newcomer China Netcom, themselves engaged in battles for control of the national service. It sought a public stock market initial public offering to generate capital, but its lackluster performance and turnover of management positions had complicated its listing hopes.

A further potential competitor for telecommunications customers, China Satcom, saw an early life that in some ways paralleled that of Railcom/Tietong. In mid-2000, China Telecom spun off its satellite communications arm to form a new company, ChinaSat. At the end of 2001, ChinaSat merged with the China Orient Telecom Satellite company and other satellite service corporations to form a unified satellite services company, China Satcom. The company did not offer residential communications services, and its main product was leasing space segments to other telecommunications operating companies and public broadcasters.[121] The company tried to expand is services when it bought a controlling stake in Unicom's paging business in mid-2004 for 710.7 million yuan ($86 million).[122] Furthermore, it held a license to operate fixed-line networks in a handful of cities, including Chongqing, Qingdao, Wuhan, and Nanjing.[123]

Despite its small role in the public telecommunications service sector, Satcom offered some hope to foreign companies still seeking to penetrate the telecommunications services market. From late 2003, the American communications company Nextel began talks with Satcom to offer services on a new 3G mobile phone network. In late 2004, Tietong joined in the discussions.[124] By the end of 2005, however, the joint venture had failed to materialize, and Satcom remained the smallest of China's telecommunications service providers.

Competition in the 1990s and 2000s

Netcom: a new competitor for fixed-line service appears

In late 2001, the Chinese government decided to induce further competition, and again split the fixed-line China Telecom company into two parts, a new China Telecom and a new China Netcom. Even with the myriad of companies that had emerged in the previous years, China's impending entry into the WTO, and potential further competition from foreign competitors, made the need to sharpen domestic services more vital. This major division of the national phone company seemed designed to prepare for WTO conditions, ones which could open the doors wider to foreign companies. The next part of this chapter gives more detail on WTO entry and its impact on telecommunications operations. But the 2001 split of China Telecom was meant to enhance the country's major fixed-line operation companies' competitive abilities against foreign challenges.

In May 2002, then, the new China Telecom took control of networks in twenty-one provinces in southern and northwestern areas, and held 70 percent of the transmission network assets. Netcom held the ten northern provinces, and also absorbed 'small' Netcom, an Internet network company with the same name. (The former Netcom is discussed in more detail in Chapter 4.)

Less than six months after its launch, Netcom was already eyeing some of the wealthier provinces in the nation's southern regions. Jiangsu Netcom was one of the first ventures to cross the boundary, and was invested with 500 million yuan ($60.2 million) in November 2002 to compete for customers in the major provincial phone market.[125] The following month, Netcom put the same amount of investment in China's largest provincial market, the southern Guangdong province.[126] Netcom pushed into Shanghai in late 2003.[127]

Not to be outdone, China Telecom moved back north. It began tentatively, setting up its Beijing Telecom branch in June 2002, but then opened in Hebei in December. By the middle of 2003, it had set up branches in the rest of Netcom's territory. China Telecom proceeded to try to hire Netcom employees for its new offices, and offered competitive prices to win business away from the northern counterpart.[128]

As the new decade proceeded, the companies found features besides fixed-line voice provision to be areas for competition. Both companies offered home modem dial-up for Internet services, and in mid-2003, China Telecom was granted a special phone number (16300) to allow customers in northern China to access its data services.[129] Both companies were also forging ahead with attracting high-speed, broadband

Internet customers. However, the physical division of the network into northern and southern backbones diminished the opportunities for competition in this arena.

As of 2007, both companies were expecting licenses to offer 3G Internet mobile services, and this would be a new field for competition between the two companies. China Mobile and China Unicom, and perhaps the two other minor companies, would also be part of this competitive field. The following chapter gives more detail on Internet services and their growing attraction to a broad range of Chinese citizens in the mid-2000s.

Impact of WTO Entry, and Expansion Outward

On 11 December 2001, after fifteen years of negotiation, China formally joined the WTO. Entry provided China with access to other nations' markets in many fields, but in return the PRC was to open many of its own sectors to foreign competition. For telecommunications, China agreed to permit up to 50 percent foreign ownership of fixed-line phone ventures beginning in 2003, and as much as 49 percent ownership of mobile ventures by the end of 2006. Fifty percent foreign ownership was allowed in 'value-added' services such as Internet access and online data processing, with several of China's large coastal cities the first to allow such cooperation.[130]

As noted above, these concessions provided some of the impetus for reorganizing the structure of the main telecommunications corporations, and for sharpening the competitive strength of domestic players. Separating China Telecom into two separate regional companies would release competitive energies and make them more formidable rivals to potential foreign entrants. However, as the early years of Unicom showed, China remained reluctant to allow in foreign competitors, even after WTO admission. As of late 2006, the only foreign company with even a minimal stake in operating a communications network was AT&T, with a limited venture in Shanghai.[131] (This case is discussed in more detail in Chapter 6.)

In theory, foreign competitors could gain entry to the market through joint ventures; but, if none of the limited universe of Chinese partners decided to team up with a foreign company, entry would be stifled. Perhaps more importantly, foreign companies likely feared the difficulty of competing with the established Chinese corporations on their home turf, with the added disadvantage of being limited to a minority stake with one of the players themselves.

One further restriction likely also limited the number of potential foreign suitors: the country set minimum financial requirements for foreign companies that did think of investing. For example, companies wanting to invest in national or cross-provincial basic telecommunications services had to have registered capital of at least 2 billion yuan (about $240 million). Small foreign companies were thereby shut out.[132]

Overall, then, in the early years of WTO membership, the nation did not face an onslaught of foreign corporations grabbing large chunks of the Chinese telecommunications market. In this sense, global competition did not materialize to further energize the domestic companies' innovation or efficiency. As noted, however, the specter of such competition had served as a useful tool to reform and reorganize the key Chinese companies in the late 1990s and early 2000s.

Despite the lack of immediate foreign competition, one other part of the WTO agreement did stand in theory to have impact on the way the telecommunications companies were regulated internally. On entry, China accepted the WTO Reference Paper, a guideline for nations that have imperfect regulatory agencies. The rule obligated China to create a truly independent regulatory body.[133] However, more than five years after entry, the MII remained essentially a tool of the one-party state, so by most standards the nation continued to lack a regulator free of political influence.

Though foreign companies had limited opportunity in China, one of the major domestic operating companies began to look abroad for new global markets. In 2005, China Mobile made an attempt to purchase 26 percent of Pakistan's top phone company, PTCL, for $1.4 billion. It lost out, perhaps for politically protectionist reasons, to a rival company from the United Arab Emirates that paid the same amount.[134] In mid-2006, China Mobile's $5.3 billion bid to buy Luxembourg-based Millicom International Cellular, which operated in Asia, Africa, and Latin America, also fell through. Millicom asserted its belief that China Mobile lacked the resources to complete the purchase.[135] However, in February 2007, China Mobile paid Millicom $460 million to buy about 89 percent of Pakistan's Paktel mobile phone company, and bought the remaining stake in April of the same year.[136]

Despite this somewhat mixed record, China Mobile's leadership noted that it would continue to seek overseas acquisitions. The company's chairman, Wang Jianzhou, stated in August 2006 that 'Our "Go Global" policy has not changed. We are still searching for opportunities abroad to grow our earnings.'[137]

Technological Advances: 3G Mobile Phone Standards and New Competitive Opportunities

As discussed above, by the middle of the first decade of the 2000s, China was poised to introduce a fully functional 3G network for mobile phone users. The service promised more rapid and complete Internet and data service access than was available with Unicom's CDMA network. As of late 2006, only about 12 percent of Chinese Internet users (discussed in more detail in Chapter 4) accessed the data network through a mobile phone.[138]

Not only the two officially licensed mobile carriers, China Mobile and Unicom, sought to benefit from the new technology—as previously noted, the major and minor fixed-line corporations also stood to become providers of the service. One issue complicating deployment of the new technology, however, was the choice of a new standard, one that would go beyond even the CDMA that Unicom had introduced for its rather rudimentary data services.

In the late 1990s, China had begun work on its own 3G standard, one that would potentially help it avoid paying royalty fees to foreign companies for using technologies developed outside of the country. The Chinese company Datang Telecom Technology worked with Germany's Siemens to introduce TD-SCDMA (Time Division Synchronous CDMA). State-owned Datang (governed by the Chinese Academy of Telecommunications Technology) and other Chinese companies expected to reap rewards if one or more of the major mobile phone companies adopted its technology, rather than use the foreign W-CDMA or CDMA-2000 variants.[139] But the two foreign standards were seen as more reliable and mature than the relatively new Chinese version.[140]

As China moved to introduce the 3G systems in advance of the August 2008 Beijing Olympics, it first appeared that the TD-SCDMA standard would win the day: in January 2006, the indigenous technology was the first approved for high-speed mobile data communication, and China Mobile allocated 26.7 billion yuan (about $3.4 billion) in March 2007, to build a trial network in eight cities. However, in May 2007, the other two standards were also approved. Other companies continued to hope for mobile phone licenses to operate 3G networks, and waited in mid-2007 to see which standard they could use. However, according to the official Chinese Xinhua News Agency, the government would wait until after the Olympics to assign these licenses.[141] With several new companies offering the next generation of mobile services, competition between the state-owned companies would become even fiercer.

Conclusion

The competition between Chinese ministries and their corporations illustrates that, even within the state sector organizations can generate the kind of dynamic that spurs both price competition and the desire to expand services quickly to the market place. Though all under the same government roof, inter-ministerial and even, after 1998, intra-ministerial rivalry was a key factor in spreading mobile phone connections in a way faster than that of nearly any other major developing country.

As Table 3.6 shows, by the middle of the decade rates for both fixed-line and mobile service had fallen to levels that were competitive with many of China's Asian neighbors. The cost relative to income of both services was about the same as that of Thailand, and significantly less than for Indonesia, India, or (for fixed lines) the Philippines. China had a long way to go to match cost levels in Singapore. However, as Chapter 7 will discuss, both fixed line and mobile services were far more affordable for wealthier urban dwellers than for the poorer rural Chinese citizens.

The central government, in the form of the State Council and its leading group, was a key factor in overseeing the telecommunications competition in the early years. It specifically recognized the virtue of breaking the

Table 3.6. Monthly price basket for fixed-line and mobile telephone use for China and selected Asian nations, 2004

Nation	Monthly price basket for fixed-line (percent of GNI per capita)	Monthly price basket for mobile phone use (percent of GNI per capita)	GNI per capita, Atlas method (current US$)
China	3.32	3.44	1,290
India	6.21	6.29	620
Indonesia	6.48	4.82	1,140
Malaysia	2.24	1.44	4,650
Philippines	12.51	4.14	1,170
Singapore	0.33	0.28	24,220
Thailand	3.94	3.23	2,540
Vietnam	9.32	15.04	550

Notes: The term 'market basket' for fixed-line residential service includes one-fifth of the installation charge, the monthly subscription charge, and the cost of 15 peak and 15 non-peak local calls. The term 'market basket' for mobile is based on the prepaid cost of 25 mobile calls on the same network and other networks, with calls including some mobile to fixed calls made during peak, off-peak, and weekends. Thirty text messages left per month are also included. GNI refers to Gross National Income. GNI takes into account all production in the domestic economy (i.e., GDP) plus the net flows of factor income (such as rents, profits, and labor income) from abroad. The Atlas Method smoothes exchange rate fluctuations by using a three-year moving average, price-adjusted conversion factor.

Source: Llewellyn Toulmin, et al., 'Telecommunications Sector: Current Status and Future Paths', World Bank, Global ICT Department (2006): 30.

Competition in the 1990s and 2000s

MPT's monopoly and providing political cover to Unicom in its early years. But reorganization of the top leaders in the late 1990s, along with the correct gamble that competition could continue even under the new MII's control, led to growth that was even more bureaucratically concentrated but, apparently, no less fierce. It is significant that the MII did not simply merge Unicom with its rival, but rather chose to continue the competitive trend under one ministerial umbrella.

The Unicom case also illustrates that a developing country like China can benefit from competitive forces even without a direct foreign challenge. Throughout the 1990s and into the first part of the next decade, the telecommunications operations sector remained basically closed—with the exception of the indirect CCF projects—to direct foreign involvement. Even within the CCF scheme, foreign companies also did not usually compete directly against each other, as they were divided up into regional markets.

Foreign companies were not only kept out of directly operating a telecommunications system, they were in the end the victims of Unicom's drive to compete with the former MPT/China Telecom monopoly. The companies unwittingly spurred Unicom's growth, but then were cast aside just as the potential for realizing a return on their efforts began to materialize.

By the first years of the new decade, the drive to create domestic competition may have gone too far. The two smallest companies, Railcom and Satcom, struggled to find market share. China Telecom and Netcom competed for fixed-line customers, but the nature of the network made it hard to secure customers in each other's territory. The spread of *xiaolingtong* offered the companies a chance to cash in on the local-limited mobile phone market, but both companies hoped to become fully licensed mobile phone operators with the advent of 3G systems.

As if to underscore the overabundance of competition, the State Council orchestrated a musical chairs shuffling of company leaders in late 2004. The president of Unicom, Wang Jianzhou,[142] became president of China Mobile; Chang Xiaobing, vice director of China Telecom, took Wang's position; and Wang Xiaochu, president of China Mobile, took over as president of China Telecom. According to a Chinese press report, 'the major reason [for the moves] is to convert rivalries into alliances and thus ease competition among them before China enters the 3G era'.[143]

As the decision to deploy 3G approached, the government may have feared wasteful equipment spending if too many companies tried to build new mobile data networks. In the year following the personnel changes, rumors indicated various combinations of consolidation and mergers among the major players. Such moves would also create stronger companies as the specter of foreign competition loomed on the (still distant) horizon.

Despite the obstacles foreign companies faced in entering telecommunications operations, the 2001 WTO agreement did open the door for potential outside involvement in the sector. Foreign company participation could unleash new forces of competition, ones to offer lower prices, better service, and new technologies. And despite the overextended nature of company splits in the middle of the new decade, the events from 1994 to 2006 show that bureaucratic competition under state ownership could bring rapid progress to a key industrial sector.

4
China's Internet and Government Policy*

From the early 1990s, Internet use in China began to grow at a tremendous pace. As of the middle of 2007, there were some 162 million Chinese with online access. Between 1998 and 2006, the country's international data bandwidth expanded by a factor of more than 3,000. In early 2007, there were nearly 2 million registered web domain names.[1]

To a great extent, the Chinese government deserved praise for rapidly building the data network and seeing that access was being granted to a quickly expanding number of the country's population. The struggle for control of cyberspace information, physical data pipelines, and network revenue, however, had a significant effect on the network's growth. The evolving demographics of Internet users and ways information was transferred within Chinese society also shaped government attitudes toward regulation of the data highway.

This chapter examines three key factors that shaped how the network is controlled in China. First, it discusses physical network control. It chronicles the ways the data pipelines came into being, and assesses the government agencies that regulated and profited from these systems. It also considers the ways the government as well as the private sector vied for control of the network infrastructure. Second, it examines network content control. Who is able to post and send information across the network, and what political limits are placed on this content? What government/private sector dynamics affect competition for web audiences, and how do revenue flows affect content-providing companies? How do user demographics determine content, and how does user reaction to content shape sociological

---

* Portions of Ch. 4 were previously published as: Eric Harwit and Duncan Clark, 'Shaping the Internet in China: Evolution of Political Control over Network Infrastructure and Content', *Asian Survey* 41 (2001): 377–408. Copyright 2001 the Regents of the University of California. Reprinted by permission of the Regents.

China's Internet and Government Policy

patterns that, in turn, influence the degree of government control over what appears on computer screens? Finally, it also considers the element of foreign influence on the network: how does foreign web content shape the Internet environment, how is it regulated, and what are prospects for change in the near future? How has China's entry to the WTO affected foreign participation in the network?

As the chapter will show, there are several areas in which these factors overlap. The result of the competing forces has shaped the ways the Internet is controlled in China, and determined what network users are allowed to see.

As in earlier chapters, we first briefly put the Chinese data network case in a comparative context.

Patterns of Data Network Construction and Government Roles

In most developed and developing nations, the government has played a key role in organizing data communication. Scientific organizations, acting with the support of public funds, were key in many countries at early stages. As economic opportunities in data transfer became apparent, private corporations soon came to play major roles in network expansion and operation.

In the United States, for example, the Defense Department's Advanced Research Projects Agency (ARPA) organized the earliest form of a large-scale data network in the 1960s.[2] The ARPANET project developed equipment to break down and reassemble data packets, and to route the data along the most efficient path. Though the system was funded and managed by the government agency, it depended on leased telephone lines from a private company, AT&T, to transmit data long distances.[3] Early network users came mainly from the academic and scientific communities.

In 1972, the ARPANET administrators had tried to convince AT&T to take over the network, so that it could expand more rapidly, but AT&T, perhaps not seeing the commercial value of data management, declined the offer.[4] The National Science Foundation (NSF) began building a parallel network in the 1980s, and in 1990 the ARPANET was transferred to management under the new NSFNET.[5] But the network infrastructure was still essentially owned and operated by a government agency. Furthermore, commercial traffic was forbidden on the government's data network.[6]

Private companies finally made their appearance in the late 1980s. Fiber optic cables offered new opportunities to move data on telephone lines, and potential commercial viability of providing data transfer became more apparent. One of the first such private companies was Performance Systems International, or PSINet, founded in 1989. PSINet bought out a regional NSFNET set of data transfer equipment in 1989. The company and other early network corporations found revenue by charging membership fees to use the system. But other private companies, such as AT&T, MCI, and Sprint, which had their own data pipelines already in place, began to offer full commercial services on parallel network infrastructure.[7]

In 1991, the NSF moved to allow full private ownership and commercial operation of its network, with Internet service providers (ISPs) allowed to operate their own backbone networks. The NSFNET was dismantled, and government ownership of the data network ceased.[8] Private service providers, such as America Online and Prodigy, which had offered limited communications opportunities among their own subscribers in the 1980s, could now offer broad access over the rapidly expanding physical network.[9] The advent of personal computers equipped with modems facilitated direct access to the data network from the comfort of one's office or home.

Over the following decade, falling access charges, along with the introduction of web browsers and search engines, and the explosion of new electronic content and services, attracted many millions of new American users to the Internet. New modes of transmitting data, such as via television cables and through wireless networks, gave affordable alternatives to leasing the telephone companies' fiber optic system. These are systems that are also becoming more commonplace in the Chinese Internet network, as is noted later in this chapter.

Other developed companies followed similar paths to expand their data network. In France, the Minitel system, opened in 1982, was entirely state-funded, and relied on the government-owned telephone network for transmission. Users received terminals for free, but paid per minute charges for access. Services were limited mainly to text, though gradually increased into greater access to the broader Internet network.[10]

Despite rapid penetration of the domestic Minitel network, personal computer-based Internet penetration remained low in France into the late 1990s, as France Telecom (FT) charged high rates for connection to private ISPs. It was only after significant public pressure that, in 2002, FT offered a flat-rate plan that lowered access charges by about 30 percent.[11]

In Japan, the private sector competed from the outset with government plans to build a data backbone.[12] In the late 1980s, universities worked, with private corporate support, to construct data networks linking campuses, research centers, and companies. But Japan's Ministry of Posts and Telecommunications was reluctant to license competitors to a slow and inefficient government-sponsored network. High costs charged by the national phone company, NTT, slowed the spread of the network. The problem of a character-based language as well as little Japanese content on the international network also slowed penetration into the 1990s. But private Japanese ISPs grew larger in the 1990s as the utility and potential of the data network became more apparent.[13]

The advent of mobile-based Internet access was key to accelerating Japanese use of the Internet. The service, backed by the semi-privatized NTT DoCoMo corporation, offered tailored Japanese language content as well as an opportunity for a society that spent long hours commuting to work or school to use the Internet at greater convenience.

For many developing countries, the data network, like the telephone system, was initially property of the state. In India, for example, Internet infrastructure was originally in the hands of a majority government-owned corporation, Videsh Sanchar Nigam Ltd. (or VSNL). The company was also India's monopoly operating company for international telecommunications services. Not until late 1999 did the government offer licenses to private (though majority Indian-owned) companies to provide Internet services.[14] VSNL itself was essentially privatized in 2002, and private ISPs proliferated. The number of Indian Internet users soared from about 90,000 in 1999 to more than 38 million at the end of 2005.[15]

Typically, government funding and guidance has played a key role in the early stages of building Internet infrastructure. Until the early 1990s, most nations did not see the commercial value of data transmission, so private companies were slow to enter the field. Once revenue streams became apparent, the government, at least in free market and politically open societies, has tended to step out of the way.

As the following sections show, the China case illustrates rapid growth even under continued state ownership of the network hardware. Moreover, Communist Party dominance has also led to a major role for the state in regulating Internet service and content provision.

Construction and Control of China's Data Network

As the last two chapters have chronicled, China's network for voice communication was constructed and controlled by the MPT and its successor MII, and later by a handful of fixed-line and mobile phone companies. The Internet network, by contrast, was largely developed in its early days by institutions outside of the MPT family. As the utility and value of data communication became more apparent, however, the MPT and its affiliated companies moved quickly to take much of the network under its wing.

As in the United States and other nations, China's first efforts at creating a data network were focused mainly on scholarly exchange of information. The country's first computer networks, the 'China Academic Network', or CANet, and the Institute of High Energy Physics (IHEP) network, in Beijing, were established in 1987.[16] China's first continuous link to the international data network came in 1988, when the CANet system established a connection with Karlsruhe University and began sending electronic mail through a gateway in Germany. At the same time, the organization chose 'cn' as the PRC's national domain name.

In the early 1990s, other educational networks arose to complement the first systems. The China Research Network, or CRnet, was established in 1990 and began by hosting more than ten research institutes. In 1992, China's leading Peking and Tsinghua universities established their PUnet and TUnet campus networks. The IHEP began using the international Transmission Control Protocol/Internet Protocol (TCP/IP) standard, with a link to Stanford University in 1994.

In the mid-1990s, the State Council decided to consolidate the disparate academic networks into one body. Following the Council's decree number 195, issued in February 1996, the CANet, CRnet, IHEP, PUnet, and TUnet were combined, under the auspices of the Chinese Academy of Sciences (CAS), to form the China Science and Technology Network (CSTNet).[17] Figure 4.1 summarizes the introduction of the various network corporations in the late 1980s and 1990s, with information about the regulating authority of the government for each.

The State Education Commission (SEC) began building its own China Education and Research Network (CERNET) in 1993. While the earlier networks were mainly regionally based in their early stages, the CERNET was planned as China's first nationwide education and research backbone with international links. CERNET's goal was to connect all of the country's universities and, later, secondary and even primary schools to one network. Early control of the data network generally fell under the auspices of the educational and academic sectors of the central government. Funding for CERNET's expansion, however, came from the larger central government budget. Perhaps more importantly, the nascent national network depended on lines leased from the MPT.[18]

In an echo of the American model, both the CERNET and CSTNet served dedicated academic and governmental customers, and did not offer commercial services to private Chinese citizens or corporations. However,

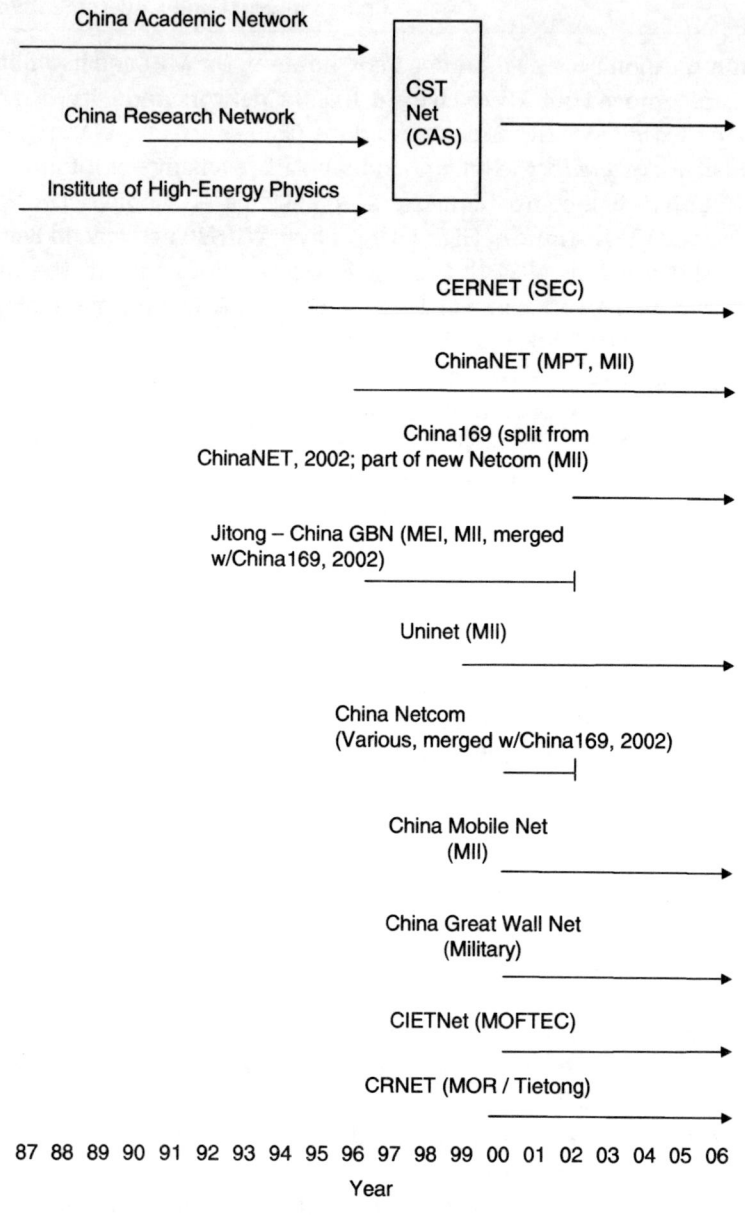

Fig. 4.1. Evolution of main academic and commercial interconnecting data networks and their regulators, 1987–2007 (from date operations began)

Note: Regulating authority is listed in parentheses. CAS = Chinese Academy of Sciences; SEC = State Education Commission; MPT = Ministry of Posts and Telecommunications; MII = Ministry of Information Industry; MEI = Ministry of Electronics Industry; MOFTEC = Ministry of Foreign Trade and Economic Cooperation; MOR = Ministry of Railways; Tietong is discussed in Ch. 3.

Sources: For information prior to 1997, see Duncan Clark, Alexandra Rehak, and Ted Dean, *The Internet in China*, Beijing: BDA (China) Ltd., 1991: 60. For information on Uninet, see China Internet Network Information Center (CNNIC), 'Statistical Survey Report on the Internet Development in China', Beijing: China Internet Network Information Center, Jan. 2000 report. For post-1997 networks, see interview with CNNIC representative, by Harwit, in Honolulu, Hawaii, 10 Jan. 2001, and China Internet Network Information Center, (CNNIC), 'Statistical Survey Report on the Internet Development in China', var. year reports.

as commercial opportunities became more manifest in Western nations in the early 1990s, the MPT itself looked to capitalize on the communications pipelines under its control.

In 1993, then, a major change in network control and development began, as the MPT started to assemble equipment to operate its own packet-data network, CHINAPAC. Newly appointed MPT minister Wu Jichuan, a lifelong telecommunications bureaucrat, had taken the ministry's helm at this important juncture, and became a leading advocate of the ministry's maintaining control in virtually all areas of voice and data communication. The MPT worked with foreign companies Sprint and AsiaInfo to design the network, and bought equipment from companies such as Cisco Systems and Sun Microsystems. The initial investment in the network was $30 million, with an additional average of $5–8 million per province for regional infrastructure.[19]

Two years later, the MPT's renamed network, ChinaNET, was launched and charged with providing public commercial services. The company was licensed as one of the government's major interconnecting networks and acted as a wholesale provider of Internet bandwidth as well as a brand name for the regional provincial telecommunications administrations (PTAs) to offer their own retail service provision. Early ChinaNET customers were to be state corporations, private companies, or wealthy individuals who could afford connection fees.[20] The network was operated and managed by the Data Communications Bureau of the MPT's communications corporation, China Telecom. Provincial branches of China Telecom managed the network services at the local level. Residential modem dial-up service was conveniently accessible and billed through the phone company, and, because of the 3-digit number customers dialed, was dubbed the '163' network.[21] ChinaNET also moved to build a parallel multimedia network, similarly named the '169' network.

The MPT soon had other ministerial rivals for control of China's data networks. For data services, the Ministry of Electronics Industry (MEI) began to compete with the MPT by creating a new corporation in late 1993. Called Jitong, it was meant to be a satellite-based telecommunications network that used the MPT's land-based telephone lines for customer local access. The new company's larger mission was to promote the so-called 'Golden Projects', ones intended to link China's customs and financial networks and provide vital information for users across the nation. Jitong's ChinaGBN (short for 'Golden Bridge Network') data web began offering commercial service in 1996. Within two years, it had some 100,000 dial-up users.[22]

The MEI project was also seen as a top-level State Council attempt to instill some competition in the telecommunications data sector. As such, the competition complemented the rivalries between China Telecom and China Unicom, as discussed in Chapter 3. As Figure 4.1 indicates, Unicom itself introduced its own data network, called Uninet, in 1999.

Despite the desire to increase competition, in the 1998 reorganization of the telecommunications sector, the MEI was merged with the new information industry ministry, which was now headed by former MPT chief Wu Jichuan. Jitong and Unicom were absorbed along with the rest of the MEI into the MII, meaning that Jitong's ChinaGBN and Uninet also came under MII authority. With the incorporation of the former State Council Information Leading Group into the new ministry, the MII emerged uncontested as the data systems' main regulator and policy director. Minister Wu, known for his fierce desire to maintain control over the course of the telecom sector, thereby also assumed a role as the PRC's information czar. He and the MII would use this control at lower levels, such as for direct provision of Internet service to business and private consumers. The growth of these service providers is discussed in a later section.

Despite this consolidation, a new rival emerged in mid-2000, as a new company, China Netcom, began operation of yet another data network. (Note that China Netcom is not to be confused with the China Telecom offshoot of the same name, created in 2002, and with which the data company merged.) The data network China Netcom company began clearly outside of MII control, with its managing partners consisting of Shanghai's municipal Information Technology Office; the Chinese Academy of Sciences; the State Administration of Radio, Film, and Television (SARFT); and the Ministry of Railways. The last, which as Chapter 3 discussed had its own fiber optic network, reportedly supplied 420 million yuan ($50.6 million) to the new corporation. Netcom's chances for success increased as China's President Jiang Zemin's son, US-educated Jiang Mianheng, vice president of the CAS and concurrently head of the Shanghai technology office, took an active interest since both of his organizations had a financial stake in the new company.[23] Netcom also received $325 million in 2001 from a coalition of foreign investors that included Goldman Sachs, Dell, and News Corporation—these companies received a 12 percent stake in the company, one that was later diluted when it merged with the 'new' Netcom.[24]

In late 2000, four more networks appeared. These included the mobile phone company China Mobile's CMNet; the military-controlled telecommunications company China Great Wall's CGWNet; the Ministry of Foreign Trade and Economic Cooperation's CIETNet; and the Ministry of Railway's CRNET (not to be confused with the CRnet, the former 'China Research Network'). The China Satellite Network (CSNet) began construction in 2001. Each of these new networks had specific target audiences: for China Mobile, wireless Internet users would be a future market, while the foreign trade network would focus on international trade via electronic commerce and the military network would be used primarily for defense-related purposes. The CRNET would utilize the railway company's rights of way to provide broadband access. Of course, each could look for larger customer bases in the future, though an already crowded field, as well as special restrictions such as those limiting the role of the military in the civilian economy, would put limits on the new entrants.

Reorganization of China Telecom at the beginning of the new decade further affected data network ownership. As noted in Chapter 3, the new China Netcom and China Telecom divided the latter company's China-NET, with Netcom taking the facilities in the north of the country, and adding Jitong and the old Netcom assets. China Netcom's services were renamed China169.

As of mid-2007, then, China had no fewer than ten interconnecting networks (though the international connections for CGWNet, CSNet, and CRNET were still in the construction stage, and CSNet did not appear to offer even domestic commercial services). Table 4.1 gives details of each network's expansion. It uses international bandwidth (the size of their connections to the international data network) as a measure of their ability to channel information.

As of 2007, the major players in running China's network were China-NET and its offshoot, China169. Both companies were nominally still controlled by the MII. With some 225,000 Mbps of bandwidth, the companies had nearly 90 percent of the nation's total international connection capacity. This allowed them near-monopoly control over China's data 'pipelines', at least for international communication.

The table indicates the educational networks remained relatively small players, though CSTNet and CERNET did continue to expand their networks at a rapid clip in the middle of the 2000s. Perhaps reflecting the rivalry between China Mobile and Unicom discussed in Chapter 3, the two companies' networks (CMNET and UniNet) held similar shares of international bandwidth as of 2007. The new entrants to the field, CIETNET, CGWNet, and CSNet, seemed to be on a slow growth track to deploy international service. The railway network, CRNET, as noted in Chapter 3, had fewer than 2 million broadband customers as of late 2005.

Though the early days of network infrastructure building in China paralleled that of other nations' focus on academic communication, commercial interests soon took a leading role in filling the data pipelines. Despite early

China's Internet and Government Policy

Table 4.1. Major networks and their leased international bandwidth (in Megabits per second) (selected years, 1998–2007)

Month/Year	Network	Bandwidth (Mbps)	% of total Bandwidth	Total bandwidth
June 1998	CSTNet	2.1	2.5	84.6 Mbps
	ChinaNET	78.0	92.2	
	CERNET	2.3	2.7	
	ChinaGBN	2.3	2.7	
June 2000	CSTNet	10	0.8	1,234 Mbps
	ChinaNET	711	57.6	
	CERNET	12	1.0	
	ChinaGBN	69	5.6	
	UniNet	55	4.5	
	China Netcom	377	30.6	
June 2003	CSTNet	55	0.3	18,599 Mbps
	ChinaNET	10,959	58.9	
	CERNET	324	1.7	
	China 169	3,465	18.6	
	UniNet	1,435	7.7	
	China Netcom	2,112	11.4	
	CIETNET	2	0.01	
	CMNET	247	1.3	
	CGWNet	(under construction)	—	
	CSNet	(under construction)	—	
January 2007	CSTNet	17,510	6.8	256,696 Mbps
	ChinaNET	135,321	52.7	
	CERNET	4,796	1.9	
	China 169	89,665	34.9	
	UniNet	3,652	1.4	
	CIETNET	2	<.01	
	CMNET	5,750	2.2	
	CGWNet	(under construction)	—	
	CSNet	(under construction)	—	
	CRNET	(under construction)	—	

Notes: Rounding numbers causes some distortion in totals. China 169 service, launched in 2003 by China Netcom, incorporates parts of the ChinaGBN network previously owned by Jitong. Jitong was merged with China Netcom in 2001.
Source: China Internet Network Information Center (CNNIC), 'Statistical Survey Report on the Internet Development in China', var. year reports.

rivals, the MII and its subordinate corporations came out as the leading operators of the country's data backbone. Furthermore, the ministry's companies used their bandwidth control and widely developed networks to garner a major portion of network user revenue, as the following section indicates.

Growth and Control of Internet Service Providers (ISPs)

Though ChinaNET and the other data networks controlled large-scale information backbones, most of the direct sales of Internet service were

China's Internet and Government Policy

left to retailing service providers. As noted above, ChinaNET in particular worked mainly as a wholesale network manager and leased its lines to provincial and other regional providers in cities across the country.

The need for a growing number of Internet service providers (ISPs) began with the explosion of Internet user numbers in the late 1990s. In 1994, there were only some 5,000 users in the entire PRC. By the end of 1996, however, there were 120,000 with access, and, as Figure 4.2 indicates, the Internet population passed 2.1 million in 1998 before reaching 8.9 million at the end of 1999, 22.5 million in 2000, 80 million in 2003, and 137 million at the end of 2006.[25] Table 4.2 puts the status in the mid-2000s in a comparative context.

China's first commercial deployment of Internet service occurred in May 1995, when the Beijing PTA, or Beijing Telecom, the capital's municipal telecommunications unit of the MPT, introduced its own ChinaNET-branded service. Shanghai's municipal PTA launched its service in June

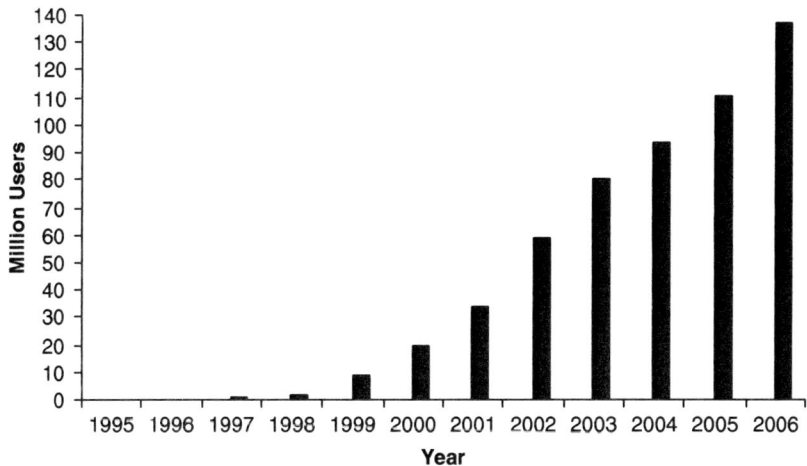

Fig. 4.2. Growth in number of Internet users, year end 1996–2006

Note: The definition of a 'user' in the CNNIC survey has varied over time. Prior to 2001, there was no clear definition that accompanied the survey results, but in the organization's Jan. 2001 survey, a user was defined as 'Chinese citizens who use the Internet at least one hour per week'. From a 2002 interview with a CNNIC official in Beijing, the definition was quoted to be 'someone who has used the Internet for at least one hour per week, on average, for the past 6 months'. Note the CNNIC compiles some of its user statistics based on landline telephone interviews, and does not conduct phone interviews with people who solely use mobile phones.

Source: China Internet Network Information Center (CNNIC), 'Statistical Survey Report on the Internet Development in China', Beijing: China Internet Network Information Center, var. year reports.

China's Internet and Government Policy

Table 4.2. Internet use and penetration for China and selected nations, 2005

Nation	Number of Internet users, 2005 (millions)	Number of Internet users per 100 Inhabitants
China	111.0	8.44
Australia	14.2	70.40
Brazil	32.1	17.24
France	26.2	43.23
India	60.0	5.44
Indonesia	16.0	7.18
Japan	66.0	51.54
Philippines	4.4	5.32
Russia	21.8	15.19
Singapore	1.7	39.79
South Africa	5.1	10.75
South Korea	33.0	68.35
Thailand	7.1	11.03
United States	197.8	66.33
Vietnam	10.7	12.72

Note: The ITU indicates different nations have different methods of categorizing 'users', so the numbers here are not strictly comparable.

Source: International Telecommunications Union (ITU) website, http://www.itu.int/ITU-D/icteye/Indicators/Indicators.aspx#.

1995 and PTAs in Guangdong, Liaoning, and Zhejiang began commercial Internet service in the second half of 1995. The first commercial service providers, then, fell clearly under the influence of the MPT's regional telephone companies.[26]

Not all of the country's service providers were local telephone companies. Accessing the Internet via bandwidth provided by the various major networks charted in Figure 4.1, other individual ISP companies, both collectively and privately owned, began to emerge from late 1995. All were at first required to obtain a license from network administrators, and many effectively operated as agents of ChinaNET. However, some ISPs did manage to develop a certain degree of independence.

InfoHighway, for example, was founded by entrepreneur Zhang Shuxin as the first private ISP in China, and began service in September 1995. The company's chief financial backer was the Hubei-based (and Peking University controlled) Xingfa Chemicals Group, which invested 80 million yuan (about $9.6 million) in September 1996.[27] Zhang sought to model her company on nascent American integrated content and service providers such as CompuServe, and targeted eight major cities for service provision. By the end of the year, there were some twenty companies, including the PTA providers, offering network connections to PRC residents. MPT-rival Jitong, at the time backed by the MEI and other ministries, launched its own ISP in September 1996, under the GBNet label.[28]

China's Internet and Government Policy

Some of the ISPs, such as ChinaNET/163 and NetChina in Beijing, were operated by or secured licenses from the local PTA and provided mainly coverage within limited localities. Others, including InfoHighway and China Online, obtained inter-provincial licenses from the MPT and attempted aggressive cross-regional or nationwide expansion. China Online, for example, sought to offer dial-up service in eighty cities across the country. Figure 4.3 charts the organizational hierarchy of Internet service provision and revenue flows.

Many of the first ISPs, however, found themselves with excess capacity and thin profit margins. These early providers began by incurring immediate losses as they spent heavily on buying or renting dial-up lines, leased

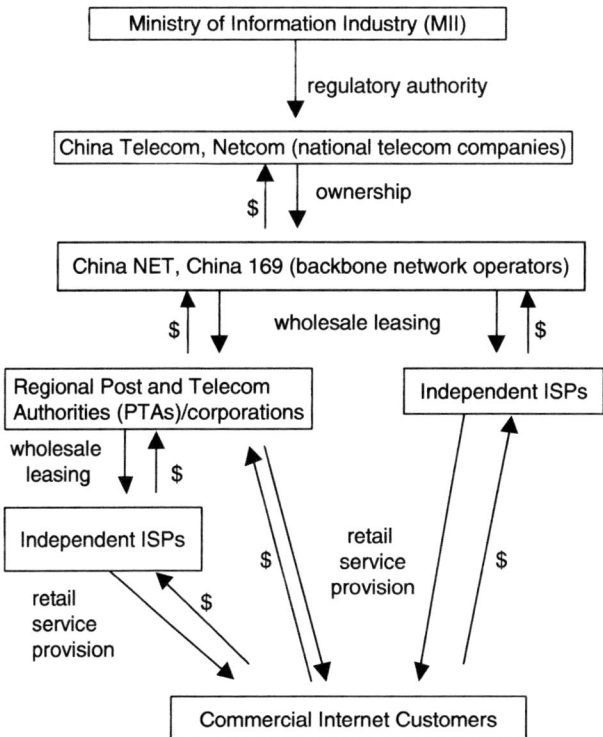

Fig. 4.3. Control hierarchy and revenue flows of Internet service provision under the MII

Notes: The '$' sign indicates revenue flow. Similar wholesale arrangements with ISPs also existed for other backbone operators such as the pre-2002 Netcom, though they fell outside of the MII purview. From mid-2000 on, almost every commercial ISP used the ChinaNET or, after 2003, China 169 backbone.

Source: Author's own compilation.

lines, Internet bandwidth, servers, and software. Furthermore, hefty online fees, averaging 400 to 600 yuan per month (about $48–$72) for 40 hours, limited customer numbers. These costs were quite high even for coastal citizens such as those in Beijing, where per capita GDP stood at only some 14,000 yuan (about $1,700).[29] Private companies, such as InfoHighway, were particularly hard hit, as they lacked the deep pockets of the government-backed telecommunications companies. Of these early providers, then, the ISPs owned by or affiliated with the local governments' PTAs emerged as dominant players. By late 1997, ChinaNET/163 in Beijing had nearly 10,000 subscribers, and NetChina about 4,000. High MPT leasing fees were also major hurdles for InfoHighway and other new ISPs. In late 1996, the ministry, via ChinaNET, charged as much as $2 million per year for a 2 Megabits-per-second (Mbps) line; in the US, the equivalent rate would have been about $500,000.[30] Leased telephone lines from China Telecom, the MII's phone company, took up to 80 percent of the ISP costs in 1999, compared to about 6 percent for Internet providers in the US.[31]

In essence, the MPT was using its near-monopoly on consumer Internet service leasing rights to draw disproportionate revenues from retailing service providers. Regional government-owned service providers could afford to wait until the number of consumers grew and profits would begin; private companies had a more difficult time. Consider the national subscriber numbers for some of China's major ISPs as of late 1999. ChinaNET-affiliated providers, such as ChinaNET/163 and MultiMedia/169, had the largest number of users, at approximately 700,000 and 500,000, respectively. The PTA-backed Capital Online also had managed to attract a large number of customers (200,000). GBNet (100,000), China Online (50,000), and InfoHighway (50,000) trailed the companies more closely associated with the MII.[32] InfoHighway's founder, Zhang Shuxin herself left the company in 1998, though internal management disagreements seem to have been the major reason for her departure. By 1999, however, the company still seemed to be doing well, with some 50,000 customers. It claimed $3 million in income for 1998, up from $1.2 million in 1997.[33]

In early 1999, a popular campaign led by academics against the high cost of Internet access and other communications caught the attention of Premier Zhu Rongji. Zhu and other central leaders addressed several of the early problems for Internet access on 1 March of that year, as they ordered sweeping cuts in leased line fees, fixed line connections charges, and Internet access rates.[34] These measures were meant both to help spread network access and to prevent the dominant MII and regional telephone company-controlled corporations from collecting excessive service fees. In the March moves, international leased line fees for ISPs were cut by some 25 percent, from $52,000 per month to $38,600 per month. At the consumer end, the price of a second line for residential users fell from $130–300 to under $30. Internet hourly rates were lowered to 4 yuan per hour (about 48 cents), bringing typical monthly bills for customers to a more affordable $15–$20.[35] By mid-2003, dial-up rates in large cities such as Beijing and Shanghai cost some 0.02 yuan (about 0.24 cents) per minute.

How did all of these price cuts affect the prospects for the smaller ISPs? The late 1990s rate reductions obviously benefited consumers, but were unfortunate for many independent ISPs: their revenue loss from reduced charges to Chinese consumers swamped any savings they could realize in the lower leased line fees they paid to use ChinaNET's backbone network. As a result, a number of independent ISPs gave up their quest for managing their own service and announced they would simply resell the service of the regionally branded ChinaNET Internet connections. By late 2000, some 85 percent to 90 percent of ISPs had such a reselling arrangement.[36]

Independent ISPs struggled for a controlling stake in the physical data network structure, but the MII and its operating companies were reluctant to allow them to collect large revenues for providing Internet access. The central telecommunications authorities and their regional carriers were able to ride out the high early costs of investing in equipment and paying leasing fees, and became the survivors while network costs fell and the numbers of users skyrocketed.

Even the independent ISP pioneer InfoHighway could not survive. After Zhang Shuxin's departure, a Hong Kong investor purchased the company in 1999. The company tried to reinvent itself as a software developer and consulting company in the following years, but in 2001 laid off most of its employees, and in 2004 lost its operations license.[37]

Still, even without the independent ISPs, new technologies and bureaucratic rivalries in the new decade helped to continue to keep costs low. Chinese citizens saw opportunities to access the Internet though high-speed data lines, mainly provided by the MII's telephone companies. In 2006, some 52 million Chinese subscribed to broadband Internet service, and about 91 million Chinese accessed the Internet through any kind of broadband connection.[38] Dedicated broadband access in the middle of the decade cost as little as 70 yuan (about $8.50) per month.[39] For corporations, a 10 Mbps line could be leased for as little as 80,000 yuan (about $10,000) per year.[40]

A new competitive force was beginning to shape the broadband market in the early 2000s. From the first years of Internet proliferation, the MII had tried to ban Internet access through cable television. However, many television stations managed to skirt these restrictions to offer competitive broadband Internet access. In return, the government's television regulator, SARFT, prohibited the telephone companies from offering television viewing through phone line connections (a service called 'IPTV', or 'Internet protocol television').[41] By 2007, the issues remained unresolved, though projections saw IPTV customer numbers growing to as many as 23 million by 2010.[42] Overall, competition between telephone and cable television providers stood to give further potential price benefits to consumers.

The result of continuing competition and government price mandates put China in relatively good stead among other regional neighbors. Table 4.3 compares Chinese citizens' monthly Internet costs in 2004 with those of several other Asian nations. As the table indicates, the monthly cost for Chinese citizens relative to GNI (Gross National Income) was rather high compared to more developed nations such as Singapore, Malaysia, and Thailand, but was significantly lower than in India or the Philippines. However, as Chapter 7 will point out, disparities between urban and rural users made Internet access far more affordable for wealthier Chinese city dwellers.

Overall, in fostering Internet service provision networks, the main priority of the MII and its operating companies seemed to have been to retain as great a control as possible over the data network, for two main reasons. First, the information ministry saw great profits by selling access on both a wholesale and retail level. Second, it saw itself as a kind of government guardian for ownership control of the communications network and its

Table 4.3. Monthly price basket for Internet use for China and selected Asian nations, 2004

Nation	Monthly price basket for Internet use (percent of GNI per capita)	GNI per capita, Atlas method (current US$)
China	9.43	1,290
India	16.92	620
Indonesia	23.43	1,140
Malaysia	2.17	4,650
Philippines	17.49	1,170
Singapore	0.55	24,220
Thailand	3.30	2,540
Vietnam	43.31	550

Notes: The term 'market basket' here refers to Internet usage of 20 hours per month, with 10 at peak and 10 at off-peak times. GNI refers to Gross National Income. GNI takes into account all production in the domestic economy (i.e., GDP) plus the net flows of factor income (such as rents, profits, and labor income) from abroad. The World Bank's Atlas Method smoothes exchange rate fluctuations by using a three-year moving average, price-adjusted conversion factor.

Source: Llewellyn Toulmin, *et al.*, 'Telecommunications Sector: Current Status and Future Paths', World Bank, Global ICT Department (2006): 30.

content, in the same way as physical and editorial control of television and newspaper companies remained in other state hands.

On the first point, moves to create competition stood to weaken the MII's position and create benefits for consumers. Furthermore, if independent service providers could give better-quality network access than the government-linked companies, there could be a reduced role for government agents in control of the network systems. On the second point, though outright ownership of the network infrastructure would probably remain in government hands for the time being, we must also consider the government's attitude toward network content. We turn to Internet content in the following two sections, considering both government desires for control, and opportunities for practical use of the network.

Internet Content: Regulation

In contrast to most of the physical network ownership and management, many of the main content providers were private or cooperative companies. In theory, the independent ownership should have given content providers greater latitude in what they could offer potential viewers of their sites. However, from the early days of public Internet access and the advent of web browsers, the government sought to control what citizens could see on their computer screens.

Of course, the national government had long put restrictions on such activities as the spread of pornography, gambling, and publication of 'counter-revolutionary' materials. The policies the government adopted for the data network therefore mirrored the security concerns over traditional media and communications already noted in Chapter 3. However, the potential for rapid and wide dissemination of politically harmful information through the Internet heightened government officials' desire for control.

As early as February 1996, when the nation hosted only some 10,000 users, police in the wealthy southern province of Guangdong required Internet users to register with the police within thirty days of opening an account with an ISP.[43] By August of that year, the government had extended the requirement to the whole country, and introduced bans on politically sensitive material and pornography. Users also had to pledge to refrain from harming 'state security'.[44]

Unlike ownership and management of the network infrastructure and service provision, control of Internet content generally fell outside of the MII's purview. It was the central State Council and top Communist Party propaganda organs that established guidelines on what material was deemed sensitive, and the public security (police) offices that enforced the restrictions. In December 1997, for example, it was deputy Public Security minister Zhu Entao who announced new rules for network content posting and usage. The ministry's regulations imposed fines of up to 15,000 yuan ($1,800) and threatened other sanctions for a new set of offenses, including 'defaming government agencies', 'splitting the nation', and leaking 'state secrets'.[45]

Rules continued to tighten in the new decade. In 2000, draft laws included provisions that anyone seeking to operate 'Internet and multimedia network services' had to apply for a license from authorities under the State Council.[46] In early October of that year, State Council decree 292 required Internet content providers (ICPs) to provide upon demand by authorities all content that appeared on their sites as well as records of users who had visited their sites for up to sixty days prior to the request. ICPs were responsible for policing their own sites for 'subversive materials'.[47] In late 2000, Anhui became the first of China's provinces to set up an 'Internet police force', but twenty other provinces and cities were also reportedly preparing such organizations.[48] The following year, Publicity Department chairman and Communist Party politburo member Ding Guan'gen pledged that the government would 'tighten control and delete "harmful" material from Internet news reporting'.[49]

The government sought further control by regulating Internet cafés, which had come to be popular and cost-effective sources for accessing the network. In the early 2000s, some 10 to 25 percent of Internet users would gain access to the network in these settings.[50] In theory, they provided users some form of anonymity, though from the late 1990s customers were required to show identification, and café managers were to keep records of users.[51] However, despite these controls, government officials periodically closed unlicensed cafés; one of the first waves came in Shanghai in 2000, when half of the city's 1,000 operations were closed for inspection.[52] (Internet cafés in Shanghai are discussed in more detail in Chapter 6.)

By the end of 2006, some 32 percent of users reported they gained access to the Internet through cafés.[53] However, the government continued its periodic sweeps of the establishments: in 2006, Guangdong province alone closed more than 70,000 'illegal' cafés.[54] By the mid-2000s, the government's main argument against the Internet bars was that they contributed to the moral corruption of youth, who used the Internet to access violent or pornographic materials, and one source indicated that 'some 80 percent of all juvenile delinquents [in China] had been addicted to the Internet before turning to crime'.[55] Of course, the regulation of cafés also reduced the opportunities for discontented members of society to post critical content in ways that they could not be identified.

In 2006, the government began using a somewhat softer tactic to remind users they were being watched: cartoon figures of police officers, nicknamed 'Jingjing' and 'Chacha' (a play on the Chinese word for police, *jingcha*) began appearing on websites in the southern city of Shenzhen to remind users to avoid posting 'harmful material and information' and conducting 'illicit activities' on the Internet. In mid-2007, the government announced these figures would be introduced across the whole nation.[56]

An implicit goal of these decrees and measures was to intimidate users into censoring their own web content. As it was technically impossible for the Chinese government to screen all domestic websites at all times, the tactic of 'killing the chicken to scare the monkeys' (publicizing punishment to intimidate the masses) was one of the few tools the authorities could use to prevent ICPs from crossing politically acceptable boundaries. Some ICPs admitted they followed the government's rules, and actively checked the content put on web pages. For example, InfoHighway's Zhang Shuxin stated in a 1996 interview that if the topics her audience addressed in discussion groups turned to be too political, 'I cut them off.'[57]

From the late 1990s, the government did prosecute several Internet offenders to set examples for the general population. One of the first casualties was dissident activist Lin Hai, who was arrested in March 1998 for sending some 30,000 e-mail addresses to an online, pro-democracy newsletter in the US. He was released in September 1999, but not before news of his crime and punishment spread rapidly through domestic chat groups and bulletin boards, and served as a warning for those who would commit similar acts.[58]

In mid-2000, in one of the first high-profile cases of prosecution for Internet activities, an activist in Sichuan province named Huang Qi was arrested for posting information on his website, www.6-4tianwang.com, about victims of the 1989 Tiananmen demonstrations. In early March 2001, Huang went on trial for 'subverting state power', a crime that carried a penalty of up to ten years in prison. In May 2003, Huang was sentenced to five years in prison, and, as of 2006, his website had become a forum for sales of Viagra and other pharmaceuticals.[59] Huang was released from prison on 5 June 2005, but was confined to his parents' home.[60]

In 2005, journalist Shi Tao was arrested for e-mailing information from a newspaper staff meeting to a democracy group based in New York. His case was notable because the Hong Kong office of Yahoo assisted the Chinese government in tracing his IP (Internet protocol) address.[61] The deeper implication of this arrest was that Chinese who felt using a foreign e-mail service would shield them from government observation now likely felt a chilling effect.

Chinese who used weblogs (or 'blogs') to express ideas critical of the nation's political system also came to be targets of scrutiny in the mid-2000s. In March 2007, for example, Zhang Jianhong was sentenced to six years in prison for publishing more than sixty articles critical of the regime on overseas websites.[62] As for domestic blogs, the government considered requiring contributors to tell blog hosts their real names and ID card numbers. These plans were dropped, at least temporarily, in May 2007, as it was difficult to verify identities of users, and domestic blog hosts protested that users would simply avoid giving their real names in any event. But the government reiterated its stance that blogs could not be used for 'slander, abuse, pornography and breaches of secrecy'.[63]

Despite these high-profile cases, by the mid-2000s potentially controversial domestic as well as foreign websites had come to play a major role on the PRC's web scene. Once China opened web access to international connections in the mid-1990s, foreign content came to be an alternative arena for the Chinese Internet community, and provided ways to skirt censorship for those seeking politically sensitive information on developments in the PRC. For example, overseas Chinese scholars founded China News Digest to spread news of the PRC derived from mainly Western news wire services. Readers could subscribe through e-mail accounts and receive daily briefings that could be displayed in Chinese language if users had proper software.

Central government control of access to foreign web pages was contradictory. As part of its larger goal of modernizing the economy and spreading scientific, economic, and cultural information to and from China, the government welcomed data network ties with foreign nations. This goal of exploiting the international network's educational and commercial advantages conflicted, however, with the desire for information monitoring.

From the mid-1990s, then, the Chinese government made several attempts to regulate access to sensitive or undesirable sites by selective blocking. Officials tried to block Western news sources sometimes critical of China, such as the *New York Times*, the *Washington Post*, and *Time* magazine. Following rules similar to those for controlling domestic ICPs, other targets for control attempts were pornographic sites, web pages printing or even broadcasting anti-government propaganda, and gambling sites. Sites promoting independence for Taiwan, Tibet, or other regions of China were also targets, as were groups promoting various strains of religious practices that lacked government sanction.

Many of the attempts to control access to foreign sites either failed or were unevenly enforced. For example, at times when *Time*/CNN websites were sometimes blocked, *Newsweek* and ABC News sites were often open. Rules announced in January 2000 that would have required web companies—domestic and foreign—that used encryption software to register each individual user were quickly withdrawn after American companies argued they were too restrictive.[64] A Harvard University study found that in 2002, some 19,000 of 200,000 sites checked on a regular basis were inaccessible inside China.[65] Still, this meant some 90 percent of the sites surveyed were not blocked.

As for skirting blocked foreign web pages, Chinese users could employ several methods to get around government blocking methods. Proxy servers located abroad, which users in China could access by reconfiguring their web browsers, provided a convenient tool to jump over what came to be called the 'Great Firewall of China'. When the Chinese government targeted the identifying numerical web addresses of offending computer servers, the server administrator could modify the number, allowing access to users who could still direct their computer to the same web address name. Another method for users with overseas accounts would be to place an international phone call to an overseas service, again allowing a jump over the international gateway-blocking mechanism.

The government seemed to be particularly tolerant of pornography. Evidence of such lax control came from one European survey firm, which found that in 1999 some 60 percent of web hits by users monitored in Beijing were adult-oriented sites.[66] The 2002 Harvard study noted above found fewer than 15 percent of pornographic websites were blocked.[67] As accessing pornography was an apolitical act, the government likely gave it lower priority for scrutiny and prosecution.

Apolitical acts also seemed to have a higher level of tolerance. On 1 July 2006, for example, Chinese taxi drivers in Beijing carried out a one-day strike to protest government failure to deliver promised subsidies. The strike was organized to a large extent through e-mail and website communication. The strike was apparently successful, as drivers soon after received the promised subsidies.[68]

Of course, the Internet is not the only tool Chinese citizens have for receiving uncensored domestic or foreign information. As earlier chapters have pointed out, Chinese citizens have access to hundreds of millions of both fixed-line and mobile telephones, as well as tens of millions of fax machines, and, in some years, as many as 30 million illegal satellite dishes.[69] These tools of communication are available to a broad segment of Chinese society and, in the short run, are probably the more likely means of communicating with disaffected members of Chinese society, at least in the near term, than is the Internet.

Be that as it may, there was some tangible evidence of foreign websites affecting domestic Chinese social behavior. In early 1999, for example, members of the Falun Gong spiritual exercise group, led by a Chinese individual in exile in the US, staged demonstrations in China that were reportedly coordinated via an American-based Chinese language web page. Such activity shows the potential power of sites located outside of Chinese government jurisdiction, and indicates regulation of domestic portals is somewhat of a moot point.

As noted above, however, the reach of Chinese censorship had come to transcend the country's borders by 2005. The case of Shi Tao indicated foreign companies were willing to assist the government in finding citizens seen to be challenging the rules of the political system. In early 2006, the American company Google announced its Google.cn site would exclude e-mail messaging and the ability to create weblogs (blogs). Google, as well as Yahoo and Microsoft, had earlier begun cooperating with the Chinese government to filter out information deemed sensitive to the authorities.[70]

The government's desire to maintain political control of the Internet remained evident as the network spread deeper into elements of the population who could use it to organize opposition movements. On the domestic scene, police control and self-censorship kept most citizens from transgressing government regulations. Intimidation of select foreign information providers further strengthened the hands of the authorities. As the next section will show, however, the need for a tight grip on politically related content did little to slow the growth of Internet content in a vast array of apolitical areas. In particular, business and entertainment were key factors fueling the continued growth of the Internet in the late 1990s and 2000s.

Internet Content: Growth in the Commercial Sector

While the government's security and ideological agencies were chiefly responsible for regulation of politically and culturally sensitive Internet content, the MII tended to concentrate more on issues related to development of the network and promotion of commercial activity. MII Minister Wu Jichuan's own statements in public venues in the late 1990s and early

China's Internet and Government Policy

2000s tended to focus purely on matters related to telecommunications business regulation and strategies for developing the industry and maintaining competitiveness in a more open international environment.[71] Wu more than likely agreed with political regulations on Internet content (discussed above), but as long as the activities of his own ministry remained relatively unaffected, he tended to avoid crossing into the realm of other parts of the government that focused on these issues.

Despite the government's early and continued desire for content control, then, the number of domain names registered in China (under the country's .cn designation) grew quickly in the early 2000s. As Figure 4.4 shows, the number of domain names increased nearly fifteen-fold from some 122,000 in early 2000 to about 1.8 million in early 2007. The country followed the pattern of other nations in corporate and private citizen domain name registration, with some 90 percent of all registered names in early 2007 labeled '.com.cn' or the generic '.cn'. Only 1.6 percent were official '.gov.cn' sites, and a mere 0.2 percent belonged to the early academic domain, '.edu.cn'.[72]

Popular domestic sites from the late 1990s included Netease (aka '163.com'), Sina.com, and Sohu.com. These sites provided mainly news, entertainment, and sports information, but often relied on officially sanctioned agencies for their own content. Some, such as Sohu.com and

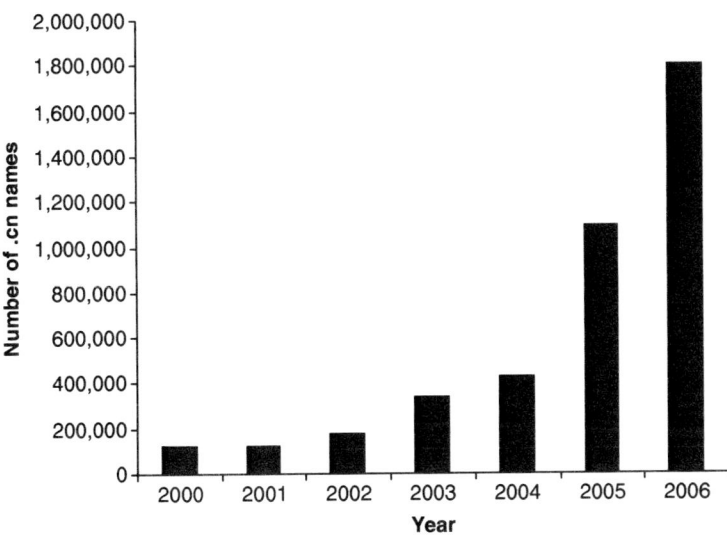

Fig. 4.4. Numbers of .cn registered domain names, year-end 2000–2006

Source: China Internet Network Information Center (CNNIC), 'Statistical Survey Report on the Internet Development in China', Beijing: China Internet Network Information Center, Jan. 2006 report, p. 34.

China's Internet and Government Policy

Table 4.4. Top web domain sites for all users, February 2001 and June 2007

Rank	Top ten in 2001	Unique users (in millions) for 2001 rank	Top five in 2007
1	sina.com.cn	4.6	baidu.com
2	sohu.com	4.5	qq.com
3	163.com (Netease)	4.0	sina.com.cn
4	chinaren.com	3.0	sohu.com
5	yahoo.com	2.3	163.com (Netease)
6	microsoft.com	2.1	
7	etang.com	2.0	
8	163.net	1.9	
9	263.net	1.6	
10	china.com	1.3	

Notes: For 2001 ranking: The IAMASIA survey was of 6,000 residential users, taken via data collection software on users' computers, across several major Chinese cities, Feb. 2001. For 2007 ranking: Alexa Internet Inc., owned by Amazon.com, analyzed web usage based on passively provided data from users of its toolbar on main affiliate websites amazon.com, archive.org, and others. Because these numbers may have some distortion for the China case, the ranking is only approximate. However, it does overlap with estimates from other sources and reports, such as 'A Web powerhouse has Beijing worried' in *International Herald Tribune*, 6 Feb. 2007, p. 15; and 'QQ Overtook Sina by Traffic' in Financial Times Information, SinoCast China Business Daily News, 15 May 2006 report.

Sources: For 2001 data: IAMASIA corporate survey, http://www.iamasia.com/; for 2007 data: Alexa Internet Inc. website, http://ww.alexa.com/site/ds/top_sites?cc=CN&ts_mode=country&lang=none.

Sina.com, provided links to foreign news about China and Chinese-language sources published in various foreign countries, but, from the early 2000s, began adopting more links to the state news organization Xinhua. Table 4.4 shows some of the pioneers in the country's Internet landscape, as well as leaders in the latter part of the decade.

Though many of the most popular early domestic websites were private, some regional government organizations tried to attract viewers with useful material. Capital Online, the web page of the Beijing municipal PTA, was popular because it offered free e-mail service. The Shanghai city government launched its Eastday.com site in 2000 to provide local news of the municipality. At the site, Shanghai audiences could view news from several of the city's main newspapers and television stations. According to Shanghai city government officials, the site became the city's most popular one within two months.[73] (Eastday is discussed in more detail in Chapter 6.)

As with the private Internet service providers discussed above, private content companies early on faced problems of generating profits. Companies that provided free content or services on their sites discovered, as did many American and other foreign companies in the Internet 'boom' years of the early 2000s, that finding paying customers could be difficult. One of the main revenue generators, online advertising, brought only $3 million to all Chinese ICPs in 1998 and some $10 million in 1999.[74]

A 1999 survey in Beijing found only three of sixty-seven Internet companies were profitable.[75] Many of the private ICPs faced looming debts in late 2000 and sought infusions of foreign venture capital.

State-owned websites could feel secure with the financial backing of a city or regional government to support them over the longer term. However, as Table 4.4 indicates, none of the state-backed websites ranked in the most popular web pages in either 2001 or 2007, indicating they were unable to attract national audiences.

By the mid-2000s, it became clear that private content providers could prosper on a national scale. Early leaders Sina and Netease reinvented themselves as providers of online games and entertainment, which users could access through the rapidly growing number of mobile phones (as discussed in Chapter 3). Latecomer Shanda Interactive, founded in 1999, also quickly grew to provide online games, and even took steps to make a hostile takeover of Sina in 2005.[76] By providing entertainment material that was difficult to counterfeit, the companies had found a new business model that sustained their revenues and profits. The share prices of most of the companies listed abroad had risen sharply in the middle of the decade as their strategies seemed to be succeeding.

In another successful move, Chinese private content providers were quick to mimic the business models of the most successful foreign companies. For example, the company Baidu, founded in 2000, learned from the success of its American counterpart Google, and surged to become China's leading search engine, with a 45 percent market share by 2005. When Baidu made a public stock offering on the American Nasdaq exchange in August 2005, its shares jumped a record fivefold. But Baidu faced competition from Google, which owned 2.6 percent of Baidu.[77]

Eachnet, which began operations in August 1999, took on the American auctioneer eBay model, styling itself at the beginning as an online flea market. As it became more successful, the company embarked on a course of buying up rivals. By 2002, the company claimed 3.5 million registered users, and one successful auction every minute. In March of that year, eBay spent $30 million to buy one-third of the company.[78] In June of the next year, eBay spent another $150 million to take control of the company.

The online auction field was not left to eBay alone. Still other Chinese private companies joined the mix in the early 2000s, including Alibaba and its subsidiary Taobao, which was a direct rival to eBay–Eachnet. Alibaba had become the largest business to business (B2B) website in China by 2002, when B2B e-commerce sales reached some 178.4 billion yuan ($21.5 billion). Consumer to consumer (C2C) sales were slower to take off, reaching only 2.5 billion yuan ($300 million) in 2002.[79] By the end of 2005, B2B sales reached 644.6 billion yuan (about $80 billion), though C2C and B2C (business to consumer) sales continued to trail, with sales of about 35 billion yuan (about $4.3 billion).[80] Chinese companies, and Alibaba in particular, reaped the benefits of this rapid commercial expansion.

Despite the rapid rise of Chinese domestic Internet companies, some of the above examples show that foreign Internet players had also begun to play a major role by the early years of the 2000s. As Chapter 3 discussed, foreign companies had been shut out of basic telecommunications operations, and the early part of this chapter showed the dominance of the MII and regional phone companies for network ownership and service provision. In the arena of Internet content provision and services, however, foreign companies found far greater areas for investment and business activity.

As Table 4.4 indicates, in the early years of Internet expansion, Yahoo's English-language page became one of the most popular sites in China. Yahoo's main attraction was that it provided a useful portal to the main American web pages. In May 1998, Yahoo launched a Chinese-language site. On the 2001 list, Microsoft and the Hong Kong-based China.com also made appearances among the most popular sites.

Following the PRC's entry to the WTO in late 2001, the country was required gradually to open up to 49 percent ownership in China's Internet service providing companies, and as much as 50 percent in other services such as e-mail and online information. Several major foreign companies quickly took advantage of new opportunities to become major players in the content arena.

For auction sites, we already noted the case of eBay's acquisition of Eachnet, a move facilitated by WTO membership. Also, as chronicled previously, Google had become active in the China search-engine market, both with its own .cn website (albeit a partially self-censored one), as well as through its part ownership in rival Baidu. In August 2005, Yahoo paid Alibaba $1 billion and handed over its China operations in exchange for a 40 percent stake in the Chinese company. Yahoo planned to expand its operations in the broad array of operations within the Alibaba portfolio, including B2B, B2C, consumer auctions, search engines, and other e-commerce activities.[81] In addition to its investment in Netcom, Rupert Murdoch's News Corporation invested a total of some $150 million in six start-up Internet companies and related telecom ventures from 1999 to 2001; but only the Netcom investment was significantly profitable.[82]

Why did the Chinese government allow such foreign participation in Internet content provision and e-commerce, while it was reticent to open the door to ISPs and other areas of telecommunications operations to outside companies? First, foreign-run ICPs were likely easier to control and less of a security risk than were foreign-operated telephone or Internet networks. As the previous chapters indicated, political and larger national security interests topped the strategic thinking of the government in its dealings with foreign corporations.

Another part of the strategy to allow foreign participation may reflect earlier trends in similar media content. In television programming for example, Joseph Chan documented the government was keen to allow restricted joint ventures in television program production so that domestic producers could both absorb needed capital, as well as learn foreign technique. He described a significant audience shift in southern China away from Hong Kong radio content and toward local production after stations actively emulated Hong Kong presentation styles using regional content.[83]

In a similar way, foreign companies could bring efficiency and quality, as well as capital, to the ICP landscape, in particular for those sites offering e-commerce services. The Chinese government would have to sacrifice some potential profits for its own companies. However, in opening China's ICP world to the outside, it would also be seen as meeting at least some of its obligations as a full WTO member.

Still, the government could continue to use policy mechanisms to keep foreign content providers from dominating the Chinese market. As Table 4.4 indicates, Chinese websites continued to hold most of the top spots in user popularity. Even a global giant such as Google, with its own Chinese-language search service, had not overcome Baidu's dominance in the field by the latter part of the decade. Part of the reason for this was continued government regulation—not until June 2007, for example, did Google receive preliminary approval to provide Internet content in China, and to advertise effectively its popular Gmail and other web services. As of mid-2007, Google took only 19 percent of the country's Internet-search revenue, while Baidu held 57 percent.[84] As with other parts of the telecommunications sector, the Chinese government would welcome a foreign challenge, but not foreign dominance.

Internet User Profiles and Social Implications

Before examining factors that will shape the development of China's Internet use into the future, we focus on ways demographics of Internet users affect both web page content, as well as possible social implications

China's Internet and Government Policy

of network viewing. The profile of Internet audiences in China helps us understand not only what material viewers will demand, but also allows us to apply larger sociological and political theories about how network users' world views will develop in the early years of the coming decade. As we will see, these world views in turn represent a feedback mechanism, one that determines the degree of social and political control that the Chinese government organs will seek to exercise.

Striking features of Chinese user profiles include the predominance of youth and male users of the data network system. Figures 4.5 and 4.6 indicate demographic patterns up to 2006. As a whole, users tended to be young men of college age. Older citizens represented a significantly smaller percentage of the network community than their proportion of the population would indicate. These characteristics reflect similar patterns to those seen in early use of the Internet in the US and Europe. As the figures indicate, though, women's usage grew from the late 1990s, but had leveled off at about 40 percent of users. Still, the closing of the gender gap reflects

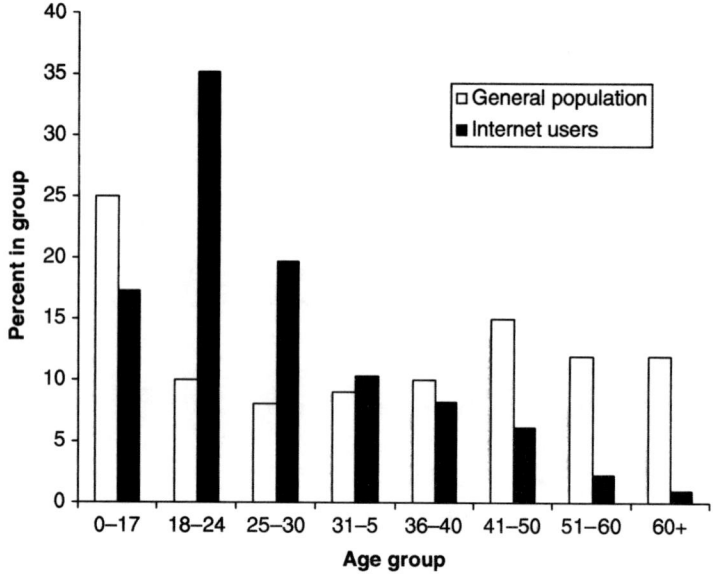

Fig. 4.5. Internet use by age, end 2006

Sources: China Internet Network Information Center (CNNIC), 'Statistical Survey Report on the Internet Development in China', Beijing: China Internet Network Information Center, Jan. 2007 report, p. 10. For general population (figure is for year 2005) see *Zhongguo tongji nianjian* (China Statistical Yearbook), Beijing: Zhongguo tongji chubanshe, 2006 CD-ROM version, chart 4–7 (percentages for age groups are approximations, as the division of the age groups were adjusted to match the CNNIC survey result age groups).

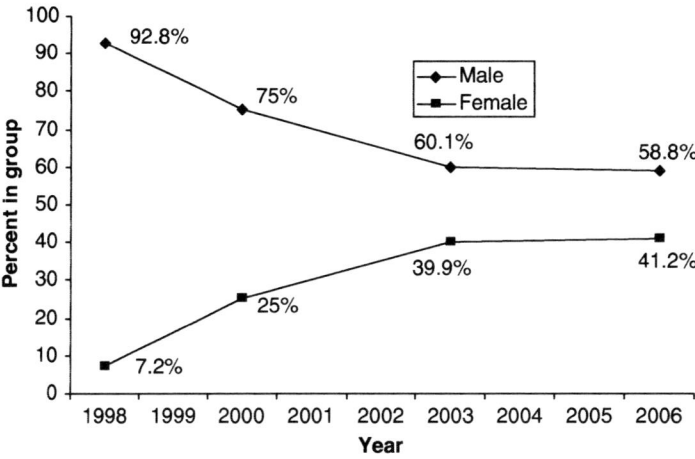

Fig. 4.6. Changing Internet use by gender, selected years, 1998–2006
Note: Figures are for mid-year.
Source: China Internet Network Information Center (CNNIC), 'Statistical Survey Report on the Internet Development in China', Beijing: China Internet Network Information Center, var. years.

patterns seen in Western countries, where age and gender features came to more closely mirror that of the larger population.

Geographically, wealthy coastal areas had a higher proportion of users than their overall populations would warrant. Chapter 7, on China's 'digital divide', provides more detail on the geographic distribution of network access. However, we note here that in the mid-2000s, cities such as Beijing and Shanghai had Internet usage several times that of poorer, inland populations, with provinces such as Yunnan, Guizhou, and Tibet seeing less penetration.

Overall, over the first decade of Internet growth in the PRC, the typical Internet user was a 20- to 30-year-old male living in either an urban area or a relatively wealthy province such as Guangdong. Noting that in the 1980s young college-age students were among the leaders of several democracy campaigns that culminated in the 1989 Tiananmen Square movement, we wonder, could or will this age cohort use new network tools for social activism purposes? In answering this question, we examine ways that Chinese citizens use the Internet and compare them to the findings of Norman Nie and Lutz Erbring, who considered Internet use and subsequent social isolation in the United States in 1999. Table 4.5 compares American and Chinese activities using web accounts.

We see that Chinese patterns are similar to those in the US and indicate utilization of the network for informational, educational, as well as

Table 4.5. Comparison of selected activities of American and Chinese Internet use patterns

Activity	American user 1999	Chinese user 2003	Chinese user 2005
E-mail	90%	91.8%	64.7%
Instant message	n/a	7.8%[a]	41.9%
General information	77%	n/a	39.8%
News	n/a	n/a	67.9%
Search engine	n/a	70.0%	65.7%
Downloading / uploading Software	n/a	43.0%	33.8%[b]
Entertainment	36%	44.9%	37%[c]
Internet games	n/a	18.2%	33.2%
Travel information	54%	6.7%	n/a
Buying/online purchase	36%	11.7%	24.5%
Job search	26%	20.3%	18.9%
Chat rooms	24%	45.4%	23.1%
Blog	n/a	n/a	14.2%
Trading stocks	7%	5.4%	14.1%[d]
Matchmaking	n/a	2.6%	3.8%[e]

[a] In the 2003 survey, the figure is for 'short message'.
[b] In the 2005 survey, the category was changed to read 'downloading / uploading files (excluding music and video)'.
[c] The category 'entertainment' was not available for 2005; the statistic presented is derived from the results for the categories 'listening /downloading music' and 'listening/downloading video'.
[d] The category was renamed 'Online financing (banking and stock trading)'.
[e] The category was renamed 'Friends/matchmaking/community club'.

Notes: Not all categories were solicited in each survey. The American penetration rate of the Internet in 1999 was about 30 percent of the population. (The number of Internet users in the US and Canada in 1999 was about 92 million; the percentage of penetration is based on a total population of the US and Canada of some 300 million. See *Boston Globe*, 18 June 1999, Economy, p. C2.) In China, the penetration rates in 2003 and 2005 were about 5 and 9 percent. These differences complicate a direct comparison, but because the polls were taken only a few years after the 'take-off' of Internet use in each country, the numbers still reflect user patterns of early adopters of the technology.

Sources: Norman Nie and Lutz Erbring, *Internet and Society: A Preliminary Report*, Stanford, Calif.: Stanford Institute for the Quantitative Study of Society, 17 Feb. 2000, p. 9; and China Internet Network Information Center (CNNIC), 'Statistical Survey Report on the Internet Development in China,' surveys for July 2003 and Jan. 2006.

entertainment purposes. There are some exceptions, such as travel information and online purchase, that reflect different levels of disposable income in each society. Furthermore, Nie and Erbring's study found social isolation increasing when Internet use was greater than ten hours per week—in their study some 15 percent of American users spent that long per week. But in mid-2000 some 53 percent of Chinese users reported spending more than ten hours per week using the network, and in mid-2003 the average accessing time was thirteen hours per week.[85] As of early 2006, the number had increased further, to 15.9 hours per week.[86] Surveys of Chinese users do not exactly replicate Nie and Erbring's queries on whether they spend less time with family and friends, but the long hours spent in front of screens indicate some similarity to American user patterns.

Subsequent studies by Nie with other researchers confirmed that time spent on the Internet comes at the expense of other social activities. Nie and D. Hillygus, for example, concluded that 'Internet use has a considerably negative impact on sociability'. Nie and Hillygus also found that 'on average, the more time spent on the Internet at home the less time spent with friends, family, and on social activities...'.[87]

Evidence from China bears out some of these assertions. As noted in an earlier section, the Chinese government closed many Internet cafés based on the claim that they contributed to the corruption of youth. The government battled what it saw as the vices of preoccupation with online games, pornography, and gambling. In early 2007, Chinese authorities indicated some 13 percent of 18 million Internet users under the age of 18 were addicted to the Internet.[88]

Should China follow the American trend, then, the likelihood of greater 'civil society' autonomous group formation in the PRC might actually be diminished. Rather than organizing movements that might fall outside of political control, Chinese citizens could end up more isolated, and less likely to challenge rule by established authorities.

Some evidence indicates there are outlets, in particular ones of a relatively apolitical nature, that could allow for online group formation. Guobin Yang, for example, points out that some environmental groups organized themselves with seemingly little government challenge. As of late 2003, some thirty-eight environmental non-governmental organizations in the PRC had websites.[89] It is unlikely, however, that the government would allow such autonomous group formation among groups with sharper political aims.

Chat groups in China allow another opportunity for communication among like-minded individuals. For Chinese, such an outlet for discussion offers a potentially powerful medium for anonymous expression of a wide variety of opinion and thought.

Chat rooms in the 1990s and early 2000s, for example, sometimes contained bursts of frustration with China's political system, and heated discussion of contentious issues. The bombing of the Chinese embassy in Yugoslavia in 1999, for example, unleashed various political debates on several websites. Following the 11 September 2001 terrorist attacks in the United States, some Chinese websites said the US deserved what it received.[90]

Despite the controversial nature of such statements, attempts scientifically to quantify such content and categorize it as 'political dialog', 'anti-government sentiment', or some other type is complicated by the difficulty of knowing whether the chat room participant actually lives in China and is thereby subject to security bureau retaliation, or whether the person is based in the US, Taiwan, or some other country.

Chinese chat groups have, on occasion, had some verifiable influence on Chinese government officials and policy. For example, Premier Zhu Rongji apparently took a more conservative stance toward improving ties with Japan during an October 2000 visit to Tokyo when chat groups expressed some hostility toward his seemingly conciliatory views.[91] In March 2001, Zhu made a televised apology to the nation after a school explosion; his action followed fierce chat room criticism of the government investigation of the incident.[92]

Perhaps the most visible effect of Internet mobilization in the first decade of the 2000s came in April 2005, when thousands of Chinese marched in anti-Japanese demonstrations in Beijing, Shanghai, and other major cities. An online petition against Japan's application to be a permanent member of the United Nations Security Council apparently sparked the marches.[93] The protests were fueled by e-mail and websites, along with mobile phone text messages with information on when and where demonstrations would be held.[94]

Despite the visible and sometimes violent nature of these marches, it is significant that the Chinese government was able to quell the movement, with which it likely had some sympathy, in a short period of time. Had the same call been for protests against the Communist Party, it is quite possible the government would have simply tracked down and arrested those posting messages calling for anti-government activity. As for mobile phone communication, the government could also shut down the whole network if it so chose.

Despite the great potential of the data network as an organizing tool for dissent, and evidence that some will take part in demonstrations that seem at least partially sanctioned by the government, for many of the young, mainly urban, male users, there may have been little incentive to endanger their future careers by discussing or engaging in political activities. As social scientists including Andrew Walder and Margaret Pearson found, those members of Chinese society who saw a secure career path within the existing political system might be reluctant to disrupt it.[95] The potential danger of random government checks of these chat groups may have sufficed to institute self-censorship among Internet users.

In sum, the avenues for greater political dialog were expanding, and as the number and demographics of users came to change in the coming years, the kinds of discussion would undoubtedly evolve. As of the first decade of the 2000s, however, there were only isolated indications that Internet forums were contributing to a greater degree of Chinese civil society. The kind of future challenge to the Communist regime from net-based autonomous group formation, and perhaps eventually democracy, had yet to materialize.

The lack of an immediate political challenge, then, influenced the methods the government employed to control both the Internet's physical infrastructure, as well as its network content. If user demographics indicated a challenge to the ruling authority would be muted, the need for control would also be softened. As access to data network tools spreads to less-privileged members of society in coming years, though, we may expect new voices could change at least some of the tone of network content. Chapter 7 will consider how the narrowing of China's 'digital divide', with greater access of the Internet to less-wealthy members of society, may shape the future course of network use.

Conclusion

In conclusion, we see that government control of the physical network was a cardinal tenet of Internet growth in the 1990s and into the 2000s. The MII came to dominate ownership of the data pipelines, and its regional phone companies crowded out private service providers. The ministry's main goal seemed to be collecting revenue from those who used the system. At the same time, the political and security officials in the Chinese government leadership saw security concerns for maintaining ownership of the nation's web resources.

The government's attitude toward Internet service provision was similar. It initially allowed private companies to operate, but the national phone companies and their regional offices soon came to dominate in the field. By the beginning of the 2000s, most service provision also came under state ownership.

On the other hand, the government allowed much of China's Internet content to lie in private hands. Strict government rules kept most citizens in line, and self-censorship was key to preventing online formation of opposition movements. As the chapter chronicled, those few who violated unspoken government limits of expression, if detected, faced severe sanctions.

The rapid growth of Internet usage and business volume, however, indicated the government was firm in its intent to see the network expand, and become an important part of the Chinese economy and society. Even with state domination of many parts of the network, after hardly more than a decade of operation, the Internet had a major impact on daily life for tens of millions of PRC citizens.

5
Building the Network: The Role of Telecommunications Equipment Companies*

Previous chapters have outlined the ways China built its telecommunications services. For these, state guidance of and competition among telephone companies and Internet corporations were instrumental to expanding voice and data transmission.

This chapter explores the way the Chinese built the nation's actual, physical network communications grid. Beginning in the mid-1980s, the government guided a transformation of the network from a creaking, thinly spread system to a world-class utility that rivals some of the world's most advanced countries.

As this chapter will show, the Chinese government played a key role in facilitating this transformation. In building the network's infrastructure since the beginning of the 1980s, it followed a multi-stage path. The government began by importing advanced foreign technology to replace aging network equipment. At the same time, however, officials also sustained production at the existing substandard, state-owned domestic equipment corporations.

The next stage saw the government skillfully attract foreign companies to partner with the state-owned enterprises, and, if the plan went well, to transfer enough technology to make the Chinese companies competitive for both domestic and perhaps foreign markets. Foreign companies were given a share of the large market, but gradually realized they were in fact sowing the seeds of their own destruction, at least as far as the Chinese domestic market was concerned.

* Portions of Ch. 5 were previously published as: Eric Harwit, 'Building China's Telecommunications Network: Industrial Policy and the Role of Chinese State-owned, Foreign and Private Domestic Enterprises', *China Quarterly* 190 (2007): 311–332. Reprinted by permission of Cambridge University Press.

A twist came at the next and latest stage, as a major private Chinese company, Huawei Technologies, came to rival the government's own state-owned choices for the telecommunications equipment market. Still, within two decades, the country had transformed its network into a well-functioning machine, while it absorbed enough technology to keep a significant portion of the market for itself. Chinese telecommunications equipment makers had even come to vie for international market share.

This chapter reflects some of the earlier findings on state-guided competition. In the case of equipment manufacture, the government encouraged competition within the state-owned sector and even between its own enterprises and those of foreign and private Chinese owners. As with other competitive forces described earlier, the policy result saw free-market energies harnessed to power the growth of an efficient, rational, and highly effective system of telecommunications equipment production and network deployment.

Equipment Manufacture during the Mao Zedong Years (1949–1976)

As Chapter 2 indicated, before 1949 there was little telecommunications equipment manufacturing capability in China. The major components of the existing network were built using imported equipment, and concentrated in cities, such as Shanghai, where there was a significant foreign presence. The Nationalist-era Chinese did have plans to open telegraph accessory manufacturing plants in the cities of Beiping (Beijing) and Hankou, but the coming war complicated these plans.[1] During World War II, the Japanese occupation forces expanded networks in areas they controlled, in particular in northeast China but also in large cities such as Beijing, Tianjin, and other parts of the country that were strategically and economically important.[2]

Following the Soviet Union's example of consolidating industrial ownership and management under central government control, post-1949 China created specialized ministries to oversee the expansion and regulation of an entirely state-owned telecommunications sector. As noted in Chapter 2, the key organization for developing and planning the network was the Ministry of Posts and Telecommunications (MPT), established in November 1949, just a month after the founding of the People's Republic.[3]

Though the MPT was the primary ministry responsible for telecommunications equipment manufacture, the First Ministry of Machine Building and its sub-bureaus also built equipment to expand the communications system.[4] With this dual structure, the Chinese did not directly imitate the American manufacturing model (discussed in Chapter 2), in which AT&T was both the communications operating company as well as the main owner of the primary equipment manufacturing company.

Both Chinese ministries worked to build equipment-manufacturing capability, with the majority of factories making telephones, transmission equipment, and other items under the MPT wing. With assistance from China's ally, the Soviet Union, Chinese engineers made good progress in the 1950s to modernize production and expand their network. The MPT simultaneously managed telecommunications equipment manufacturers in cities such as Shanghai, Xian, Beijing, and other parts of the country.[5]

A breakthrough came in 1957, when, with Russian assistance, the Machine Building Ministry's Beijing Wire Communications Plant began production of automated telephone central office switches. These devices can be thought of as the 'brains' of the network that direct voice or data communications from one source to another. Their domestic production meant that China could more rapidly add to the capacity its communications network could carry.

Each switching machine could handle a limited number of phone calls coming across the telecommunications lines. The capacity of the switches is therefore measured in the number of lines it can manage—the unit of capacity is the 'port'. In 1949, for example, the total capacity of the country's switching network was only 312,000 ports, but by 1960 it was more than 2.4 million.[6] Figure 5.1 shows the growth of domestic switch installation in the PRC in the first four decades since 1949.

In the late 1950s and 1960s, Shanghai became a center of equipment manufacturing, and hosted both a telecommunications equipment factory and a telecommunications research institute.[7] Equipment factories emerged in many other Chinese municipalities, with Changchun, Chongqing, Tianjin, and most other large municipalities joining the above-mentioned cities in the sector. By the end of the 1970s, there were twenty-seven equipment factories under direct control of the MPT, and 100 other factories regulated at the provincial telecommunications bureau level.[8] In addition to switches, these companies produced equipment to enhance transmission as well as more basic items such as cable and the telephones themselves.

During the 1950s, the Chinese government adopted the Soviet Union's planning system for managing its production units.[9] The MPT would first set a quota, and send it down to the provincial or municipal telecommunications authority office. The local telecommunications administrators

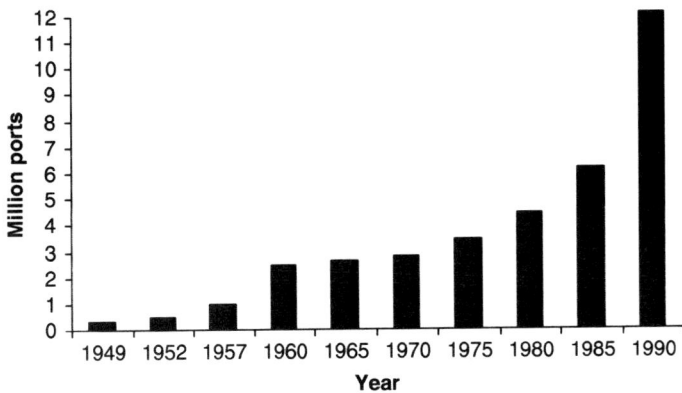

Fig. 5.1. Central office switch capacity (in total number of phone ports), 1949–1990
Sources: *Zhongguo jiaotong nianjian* (China Communications Yearbook), Beijing: Zhongguo jiaotong nianjian she, 1986: 235, and *China Posts and Telecommunications Annual Report*, Beijing: China Posts and Telecommunications Editorial Group of MPT, 1997: 42.

would make a plan for production, then send it back up to the MPT, which would then send it further up to the national State Planning Commission (SPC) for review and auditing. The final version of the plan would then be sent down again to the factory as a production plan.

As Figure 5.1 shows, the Chinese network capacity increased some eightfold by the early 1960s. The MPT and SPC were able to guide the fledging sector to a significant measure of success in the country's first decade. Moreover, under Mao's egalitarian policies, the spread of the network was not limited to the wealthier urban areas. While the number of city dwellers with phones tripled from 1949 to nearly 800,000 in 1965, rural users saw phone penetration grow more than ten fold over the same period, to about a half-million subscribers.[10]

During the Cultural Revolution years of 1966–76, when Mao directed Red Guards to find enemies of the revolution, much of industrial production was disrupted. Switching capacity barely grew over the decade, as Figure 5.1 indicates. As the worst of the chaos subsided in the early 1970s, the MPT resumed control over planning and regulation of the telecommunications sector.

The Reform of Equipment Manufacture in the Post-Mao Years

A key development in the reform of telecommunications equipment manufacture came on 1 April 1980, when the MPT created the China

Posts and Telecommunications Industry Corporation (PTIC).[11] This was an important step, as it added an element of free market corporate planning to the equipment-manufacturing sector. The new state-owned company came to be the early focal point for strengthening domestic telecommunications equipment manufacture, as it provided a channel for both corporate innovation as well as foreign investor participation in the first years of the 1980s economic revitalization process.

The PTIC was a holding company that directed all of the regional equipment manufacturing corporations. The company took over the MPT's responsibility for managing the twenty-seven centrally controlled corporations. It directed product planning, investment, research and development, capital formation, and relations with foreign companies. It could set product prices and control imports and exports. It could also both establish and terminate companies under its umbrella. For their part, the companies could decide their own marketing strategies, do some product planning, set local wages, and make other micro-level decisions.[12]

The rest of the country's 100 or so telecommunications equipment manufacturers did not get complete independence, as they were still under the direction of the provincial MPT management offices. The companies were allowed to sign business contracts as independent legal entities, but the MPT still chose the companies' managers. The PTIC was also empowered to coordinate production between the local companies and their own centrally owned factories.[13]

Despite its nominal independence, the PTIC was still seen as part of 'one family' with the MPT, and the company remained administratively subordinate to the ministry.[14] However, the new structure opened the door for the PTIC to function along free market lines: most important, its factories now had the ability to retain profits.[15] This factor inspired the PTIC to develop new products, pay attention to quality, and actively compete with rival, non-PTIC equipment companies.

In early 1982, MPT minister Wen Minsheng announced expansion of the country's telecommunications system would be a priority industry for developing the nation's economy.[16] He acknowledged that to achieve this goal, China would need to import foreign technology and equipment.[17] This aim was reinforced in 1984 by Wen's successor, Yang Taifang, who noted that the reforming economy overall was developing faster than the telecommunications sector.[18]

The early 1980s was marked by large-scale efforts to import advanced telecommunications equipment. Fuzhou, Shanghai, and other coastal cities, ones meant to attract foreign investment in a wide array of industries, were among the first to import modern digital switching and transmission systems to provide quickly the kind of networks demanded by foreign business people.[19]

Telecommunications imports got a key boost in 1986, when the State Council and other top government planning commissions and ministries approved lower tariffs for imported telecommunications equipment. According to the policy, import duties would be halved for direct imports, and would be waived entirely for telecommunications companies using World Bank or Asian Development Bank loans to buy the latest technology. This policy lasted until 1996.[20] Through the latter part of the decade, China came to rely almost completely on foreign imports for installation of advanced program-controlled central office switching equipment in its major cities.[21] The Japanese companies Fujitsu and NEC, as well as France's Alcatel and Sweden's Ericsson, came to be major suppliers of the Chinese imported equipment.[22]

The PTIC also began to play a role in importing technology and using it for domestic production. In 1984, for example, the company imported equipment to start advanced telephone cable production in the western city of Chengdu.[23] Gradually, however, the PTIC began to absorb the foreign technology, and by 1987 was already producing equipment such as microwave communications systems and optical fiber digital devices that it previously was importing.[24]

Still, importation of critical devices such as digital central office switches was expensive: in the late 1980s, the cost could reach as high as $300–500 per port.[25] The purchases began to be a drain on the central government's supply of dollars, yen, and other foreign exchange reserves. China decided to meet this challenge, as well as accelerate the drive to technology absorption, by inviting foreign companies to set up joint-venture equipment-manufacturing facilities with domestic companies.

Foreign Telecommunications Equipment Investment in the 1980s and 1990s

As noted at the beginning of this chapter, the second strategic policy path was to encourage foreign companies to set up jointly owned companies. In doing so, the Chinese could maintain managerial control and ownership rights while achieving their goal of producing modern equipment within the country's borders. At the same time, they could absorb technology to foster their own indigenous manufacturing capability. Work by Margaret

Pearson chronicled the government's ability in the 1980s to maintain control over foreign joint ventures while they received the benefits of capital infusion and technological and managerial innovation.[26]

As early as 1977, the Chinese government started talking with companies from Western Europe and Japan about cooperation in building the infrastructure for a modern communications network. American companies, including AT&T, joined the competition for participation in China in 1979, after the US switched its recognition that year from Taiwan to the PRC.

In 1983, after nearly three years of negotiations, the PTIC took on as its first joint-venture partner the ITT Corporation's Belgian subsidiary, BTM. In July of that year, the new company, named Shanghai Bell, was established. PTIC took 60 percent of the venture, BTM held 31.65 percent, and the Belgian government's Fund for Development Corporation acquired the remaining 8.35 percent. The Chinese government took majority ownership in the new company to guarantee both managerial control as well as access to the technology for making large-capacity digital switches. To further the goal of technology acquisition, the Chinese insisted that the venture should produce state-of-the-art technology.[27]

For ITT, the attraction was obvious: the PTIC was the largest equipment manufacturer in China, and its ownership by the MPT meant the company would have a nearly guaranteed market. The MPT not only indirectly ran the equipment company, it also controlled China's telecommunications operations services. In practice, according to Shanghai Bell officials, the Chinese partner was effectively the joint venture's customer.[28]

The joint venture company also received other preferential treatment. For example, it had no limit on its production, though other factories in China did have such caps.[29] To jump-start domestic purchases of the equipment, the MPT provided some 60 million yuan (about $17 million) in subsidies to domestic users who bought equipment from the joint venture.[30] MPT minister Yang Taifang announced the company's 'System 12' switches would be the country's principal source for telecommunications exchanges. Production began in 1985, and in 1986 the company produced equipment with capacity of 66,000 ports.[31] Still, this was only about 11 percent of the nearly 600,000 ports installed that year; imports continued to take the majority of market share.

In 1987, Alcatel of France took control of BTM, and inherited the stake in Shanghai Bell. Production of switches increased at an annual clip of 50 percent from 1988 to 1990.[32] As Table 5.1 indicates, by the end of the decade, Shanghai Bell (the sole joint-venture producer at the time) had about a quarter of the central office switch market. The company also was

Building the Network

Table 5.1. Central office switch sales, selected years, 1990–2003, in millions of ports

Year	Imports	Joint ventures[a]	Local manufacturers	Total installed	Total national switching capacity
1990	1.4 (71%)	0.45 (23%)	< 5%	2.0	12.3
1992	2.4 (57%)	1.6 (38%)	0.25 (6%)	4.2	19.2
1994	9.6 (51%)	7.3 (39%)	2.0 (11%)	18.9	49.3
1996	3.3 (16%)	9–10 (40–50%)	5–6 (25–30%)	20.9	92.9
1998	n/a (near 0%)	12 (46%)	14 (54%)	25.5	138
2000	0%	14 (57%)	11 (43%)	24.8	178
2001	—	29 (38%)	48 (62%)	77.4	256
2003	—	20 (30%)	45 (70%)	64.3	351

[a]Joint ventures include Shanghai Bell (for 1990–2003), Siemens BISC (for 1991–2003), Tianjin NEC (for 1992–1994), and other minor joint venture production in the 1990s.

Note: Variation in sources makes numbers not strictly comparable between years.

Sources: For 1990–2003 total installed and national switching capacity: *Zhongguo tongji nianjian* (China Statistical Yearbook), 2004, p. 662; for 1990–1994 imports, joint ventures, and local manufacturers: Zixiang Tan, 'The Outlook for the Future of China's CO Switch Market', *Telecom Asia* 7 (1996): 45 (Table 2, with import numbers adjusted to conform with total installed capacity), and *Business China*, Economist Intelligence Unit, 30 Sept. 1996, p. 8; for 1996 imports, joint ventures, and local manufacturers: *China Posts and Telecommunications Annual Report*, 1997: 39; Xinhua News Agency, 19 June 1996 report; *The Economist*, 27 June 1998, p. 65; and July 2002 interview at Shanghai Bell; and author's estimate for BISC and NEC production; for 1998 joint ventures and local manufacturers: *Asia Pulse*, 9 Apr. 1999 report; for 2000 joint ventures and local manufacturers: Alex Zixiang Tan, 'Product Cycle Theory and Telecommunications Industry', in *Telecommunications Policy*, 26 (2002): 24; and July 2002 interview at Shanghai Bell; for 2001 joint ventures and local manufacturers: *South China Morning Post*, 21 Oct. 2002, Business Post p. 4; and July 2002 interview at Shanghai Bell; for 2003 joint ventures and local manufacturers: *Financial Times Information*, SinoCast News Wire, 21 July 2003 report; *Financial Times Information*, SinoCast News Wire, 25 July 2003 report; August 2004 interview with Siemens in Beijing; and author's estimate.

widely praised within China, and was consistently ranked in the government's 'top ten' lists of superior joint ventures.[33]

As Shanghai Bell expanded production, it moved to increase the percentage of local Chinese parts in its products. The Chinese government encouraged such moves, as they would help reduce reliance on imports, and improve the country's trade balance. Localization of production would also further the goal of technology transfer. For Shanghai Bell, the company could also reduce costs if a domestic manufacturer could provide quality parts at a cost lower than that of imports. To further the localization goal, Shanghai Bell formed a subsidiary parts supplier, Shanghai Belling, in 1988, to manufacture integrated circuits for the parent company's main products. The subsidiary helped raise local parts content from 20 percent in 1988 to 68 percent in 1995.[34]

The Chinese government saw introduction of a joint venture competitor as an important way both to reduce imported equipment levels, as well as prevent Shanghai Bell from becoming too dominant among domestic producers. The company was therefore joined in the late 1980s by another major Sino-foreign joint venture, this time with Germany's Siemens AG. The Chinese had begun negotiations in 1985, just as Shanghai Bell was beginning production. In 1988, Siemens joined the switch competition in taking a 40 percent ownership stake in the Beijing International Switching Company, or BISC.

The Chinese partners for BISC clearly established it as a rival to the PTIC. The MPT was represented with its local Beijing Telecommunications Authority taking only an 8.2 percent stake. Meanwhile, the rival Ministry of Electronics Industry (or MEI, a successor to the Ministry of Machine Building) held 25 percent of the company through the Beijing Wire Communications Plant, the same company that had developed China's earliest switching systems in the 1950s.[35] The rest of the company was held by the municipally owned Beijing Comprehensive Investment Corporation.

BISC showed that not only did the government see competition between rival foreign companies as a key to improving production, it also saw inter-ministerial competition as useful to prevent the kind of bureaucratic stagnation possible if only one ministry controlled the entire sector. As in the case of China Telecom vs. China Unicom (discussed in Chapter 3), the Chinese leaders allowed and encouraged rival ministries and their subsidiary companies to compete for domestic market share.

By the end of the 1980s, the Chinese government had decided to limit the number of foreign companies that could take part in switch production. While competitiveness should be ensured, the government did not want to see excess capacity and joint ventures that could find no market for their products. In 1989, the national State Council promulgated its Directive 56, limiting foreign participation in switching equipment manufacture to three companies: Alcatel, Siemens, and NEC (still in negotiation at the time) of Japan.[36] This pattern also followed the Chinese strategy of pitting foreign corporations, often from different nations, against each other to drive down prices and raise quality while avoiding collusion among the foreign companies.[37] Moreover, the directive was meant to minimize the incompatibility of equipment purchases from a broad array of different foreign manufacturers, and to assist in standardization across regions.

NEC joined Alcatel and Siemens in 1990, as it created its Tianjin NEC joint venture in the coastal Chinese city. As with the other two companies, the Japanese held a minority share: NEC took 35 percent, and its partner Sumitomo held 5 percent. The MPT was represented through its local Tianjin Telecommunications Authority holding 15 percent, and another MEI subsidiary, Tianjin Zhonghuan Computer company, took the remaining 45 percent.

As Table 5.1 shows, by 1990 imports continued to take a large share of the market, as decentralization had allowed provinces and municipalities to choose overseas suppliers over the domestic makers.[38] However, by the mid-1990s, the central government was able to stifle imports as the domestic joint ventures and, as discussed below, local manufacturers, picked up market share.

Shanghai Bell strengthened its hand in the 1990s, as financial assistance from the French government provided necessary foreign currency for the venture and allowed it to expand its reach within the country. 'A French soft loan, and we got Beijing, Hunan, and Heilongjiang [provinces]', said Maurice Vallat, a senior Asia-based Alcatel official in 1994.[39] Production at the company soared in the first part of the decade, as Table 5.1 indicates. The company became profitable, with average annual returns of 100 million yuan (approximately $20 million) by 1991.[40] Profit for 1992 was about $70 million.[41] The People's Construction Bank of China rewarded the company that year with a loan of $50 million, and the venture's contract, originally due to expire in 1998, was extended to the year 2013. The company also began work on a new factory site in the new Pudong industrial zone of Shanghai.[42]

BISC was able to keep a significant market share of as much as 20 percent in the late 1990s, though it was perpetually trailing its French counterpart.[43] Tianjin NEC began to fade from the mid-1990s, as the weakness of Japan's domestic market put a strain on NEC's overseas operations. Still, the competition in the sector helped lower domestically produced switch prices from about $180 in 1990 to some $100–120 per port in 1995.[44]

By the mid-1990s, Shanghai Bell's switches contained about 70 percent of locally made Chinese content, as measured by price. This development pleased the Chinese government, but protection of intellectual property became a concern for the joint venture. A 1996 study by the Washington-based Center for Strategic and International Studies asserted that Shanghai Bell 'faced the problem of illicit or unauthorized technology acquisition'.[45]

Moreover, it was not just foreign companies prospering from the success of the joint ventures: the Chinese partners under the MPT and MEI (including the PTIC and other partners) were also seeing greatly increased cash flow from the booming sales of telecommunications equipment. In effect, the Chinese government was using market forces to channel revenue into the state-owned equipment manufacturers. The rise in the number of fixed-line customers among private home users, as illustrated in Figure 5.2, was the main source of the sector's revenue.

The third stage of China's telecommunications equipment policy saw the PTIC and other domestic companies try to use their growing financial

Building the Network

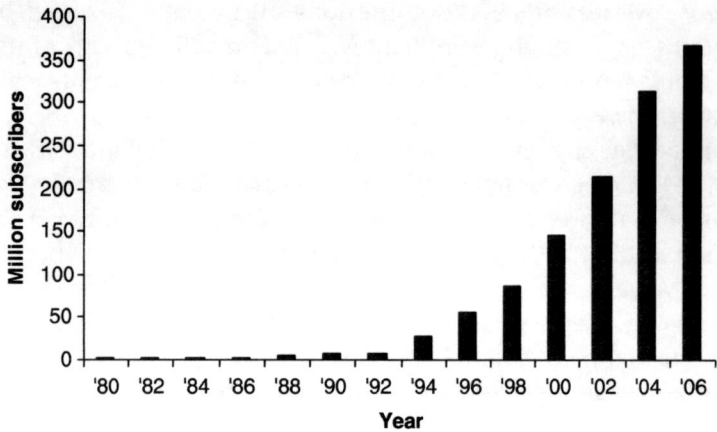

Fig. 5.2. Number of fixed-line telephone subscribers, 1980–2006
Sources: China Posts and Telecommunications Annual Report, Beijing: Zhongguo jiaotong nianjian she, 1997, p. 50; *Zhongguo tongji nianjian* (China Statistical Yearbook), Beijing: Zhongguo tongji chubanshe, 2004: 660; for 2004. Ministry of Information Industry (MII) website, http://www.mii.gov.cn/mii/hyzw/tongji\yb\tongjiyuebao200412.htm; for 2006: MII website, http://www.mii.gov.cn/art/2007/01/22/art_166_28236.html.

might to wrest control of the switch market from the Sino–foreign joint ventures. As the following section shows, the goal of fostering domestic rivals was achieved to a large degree, though ironically the Chinese private sector, in the form of the Huawei Technologies corporation, proved more adept than did the state-backed corporations.

The Challenge from Chinese Companies: Great Dragon, ZTE, and Huawei

Great Dragon—failed conglomerate, failed policy

One of the first organized government efforts to take market share from the joint venture companies was the formation of the Great Dragon Group, or 'Julong' in Chinese. The company began in late 1989 as a research endeavor at a military university, the Zhengzhou College of Information Engineering in Henan Province.[46] The project, to develop an advanced indigenous switch, was led by Professor Wu Jiangxing, the son of a People's Liberation Army (PLA) general, and financed by his college and the PTIC. The company began with seventeen people and financing of some 3–4 million yuan (about $800,000 at the time), and introduced its HJD-04 switch in 1991. The research group licensed the technology to six MPT and MEI companies in late 1991. Among these six was the Beijing Wire Communications Plant, the same company that had partnered in 1988 with Siemens to form BISC.

By1994, switch sales from these factories totaled some 2 million ports (see Table 5.1), nearly all of the solely Chinese-owned production, and about 10 percent of the nation's market that year. The cost of the switch was some 450–500 yuan per port or about $55, at a time when imported and joint venture switches, such as those made by Shanghai Bell, cost $100 or more per port.

In 1995, the Chinese government decided to amalgamate the largest domestic switch producers into one manufacturer. The Great Dragon company was therefore incorporated in March 1995 by grouping the now eight companies that were manufacturing the HJD-04 switch—among these were four companies owned by PTIC, three from the MEI, and one owned by the military.[47] Wu Jiangxing represented the Zhengzhou College group and took the position of company chairman. On the company's founding, chairman Wu noted that his corporation would rank second in the nation, behind Shanghai Bell, with production of 3.5 million ports.[48] Great Dragon set up its headquarters at the Beijing Wire Communications Plant, putting it virtually under the same roof as the Siemens joint venture competitor.

The central government extended its support to the new company with a promise of financial assistance. The government pledged an annual credit of $250 million to Great Dragon, along with $2.5 million per year for research and development projects. At the time, an MEI telecommunications engineering official, Liu Dingzhuan, complained that 'Foreign companies are flooding in. They are busy biting into big profits, while their domestic counterparts are waiting for the government's preferential policies.' Chairman Wu reflected this sentiment, saying that foreign-backed companies could thrive in China because of their own governments' soft loan policies.[49] Wu indicated that the company's goal was to take control of the market back from foreign corporations.[50]

The new company's other ambition was to lessen competition among the domestic makers who were seen as undercutting each other's prices. However, over the next few years, price wars remained a hallmark of the industry's production. The next sections of this chapter discuss how the rise of new domestic companies Huawei and Zhongxing Telecommunications Equipment Company (or ZTE) maintained price pressure on the equipment. For Great Dragon, however, the results were problematic.

By the end of 1997, the price per port of switching equipment fell to 400 yuan (about $48), and in early 1999 it was as low as 200–250 yuan

($24–30). Great Dragon chair Wu asserted that year that 'the development of the same product is something that should be avoided as much as possible.' He suggested the government should implement 'policy adjustments' so that the major domestic companies would avoid direct competition.[51]

It is significant that at this point the government refused to back down from its policy of encouraging the kind of competition that would lead to lower prices and better-quality products in the telecommunications equipment sector. The government had also probably lost faith in its first creation, Great Dragon, and had shifted its attention to the more likely winners in the domestic arena, Huawei and ZTE.

Great Dragon's problems intensified as Huawei and ZTE moved aggressively to take market share in the last years of the 1990s and first years of the new decade. Great Dragon tried to expand into wireless equipment manufacturer when it signed a licensing agreement with the American company Qualcomm in 2001, but had little success with this endeavor.

A key problem for Great Dragon was that it was created by combining eight disparate companies from two different ministries and the military. The PTIC had competing loyalties, as it had stakes in both Shanghai Bell, the market leader, and Great Dragon. Likewise, the MEI had put its money on both Great Dragon and Siemens BISC. Even following the merger of the MEI and MPT in 1998, the company faced problems of integrating its research, production, and marketing divisions.[52] The company's physical proximity to the successful Siemens joint venture seemed to be of little value.

By 2002, Wu Jiangxing was out as company chairman, and the next year Great Dragon set out to revamp its organizational framework. As of 2005, the company seemed to be struggling to find its place in the equipment arena.

Great Dragon indicates that even with central government efforts to concentrate production in competent state-owned hands, officials had backed a losing venture, and the results were a failed effort at developing indigenous telecommunications switch production. However, the government's emphasis on fostering competition between both joint ventures and domestic corporations continued during the 1990s. Two rival companies to Great Dragon, ZTE and Huawei, represented more successful ways corporations and the government worked together to bring switch production into local hands.

ZTE—the state's success story

ZTE was established in 1985, and was under the jurisdiction of yet another Chinese industrial ministry, in this case the Ministry of Aerospace Industry

(MAI). It was originally a semiconductor producer, and had as other shareholders the local Shenzhen municipal Changcheng Industrial Company and the Yunxing Electronic Trading Company.[53]

As the Chinese government saw the need for more domestic companies to produce telecommunications switches, ZTE moved in the late 1980s to develop a product to compete with Shanghai Bell and imported products. The MPT approved of the company's first switching device, the ZX500, in 1990, and the company used continued support by the MAI to transform itself into a telecommunications equipment manufacturer in the early 1990s.[54]

In the next few years, ZTE strengthened its telecommunications development skills by establishing research and development facilities in Nanjing and Shanghai. In late 1997, ZTE became the first Chinese telecommunications company to raise money through a public stock exchange listing, as it sold stock on the domestic Shenzhen exchange. As of 2004, 49 percent of the company was in stockholders hands, but the state continued to hold a majority 51 percent stake.[55]

Though Great Dragon had a commanding lead among domestic producers, by 1996 ZTE and its neighbor in the special economic zone of Shenzhen, Huawei, were producing a total of some 3–4 million ports of switching equipment, and in 1998 ZTE alone sold 5–6 million ports, about 20 percent of the national total and close to twice what Great Dragon made that year.[56] As noted above, by the beginning of the new millennium, ZTE and Huawei had essentially driven Great Dragon from the marketplace for switching devices.

Why did ZTE enjoy such success? One reason was that, unlike Great Dragon, ZTE had more focused leadership in its early years under the MAI; it did not have to answer to managers from multiple ministries, as did Great Dragon. Furthermore, after the public listing, ZTE management was no longer under government control, and management style was based on private enterprise principles. The company was essentially 'state-owned, privately managed' or '*guoyou siying*' in the Chinese terminology.[57]

According to these principles, then, ZTE used two strategies to take market share. First, it contributed to the falling prices of central office switches by offering its products at costs nearly half of those offered by the joint ventures. Second, it targeted rural markets in inland parts of the country. These areas were among the poorest parts of the nation, so were unable to afford the prices of the more sophisticated joint venture manufacturers; but products from ZTE were more suitable to their more limited needs. It was only in 1996 that ZTE even began to target larger cities for sales.[58] The company's overseas expansion plans reflected the domestic strategy, as it began exporting its switching products to developing nations such as Pakistan, Bangladesh, and Kenya in 1997.[59]

Despite the company's investment in research and development, leakage of technology from the joint ventures may have been a further factor in helping ZTE (as well as Huawei, discussed below) advance so quickly. As noted above, Shanghai Bell had begun to see a loss of its intellectual property as early as 1995, and the local companies were likely beneficiaries of this trend.

Government protectionism also came to the aid of ZTE in the late 1990s. In November 1998, the Ministry of Information Industry (MII), the successor ministry to the MPT and MEI, announced that rapidly growing mobile phone companies should buy 'local' equipment wherever possible.[60] It was unclear whether this order included domestic joint ventures, and, in any event, the edict did not directly affect the sales of fixed-line central office switches. However, by this time ZTE, Huawei, and other companies were supplementing their fixed-line equipment production with mobile telephone equipment, and the edict was a natural boon to their larger revenue streams. One Huawei official indicated that 'this is good news for us', and telecommunications analyst Duncan Clark noted that 'there will be a clear preference where possible to buy from the emerging domestic champions'.[61] ZTE took about 17 percent of the domestic market that year.[62]

ZTE's continued progress in the early 2000s is discussed in a later section. However, the corporation clearly prospered as a state-owned company in the 1990s, with notable government support to provide needed telecommunications equipment to developing parts of the nation. But the competitive environment also included the private company Huawei, and the state's industrial policy allowed the growth of this competitor to further sharpen the nation's equipment production skills.

Huawei—a prominent role for the private sector

Huawei Technologies was founded in 1988 by Ren Zhengfei, a man who played a key role in leading the company over its first two decades. Ren was born in 1944, and attended middle school in Guizhou province and university in the inland city of Chongqing. On graduation, he joined a military research center in Mianyang, Sichuan. Just after the Cultural Revolution, in 1978, he won a prize for technical achievement. Over the next decade, he sold televisions and other consumer electronics in an enterprise that lost money. Ren and his co-founders achieved greater success after they established Huawei in Shenzhen using 20,000 yuan (about $5,400 at the time).[63]

As of November 2002, Ren and his chief co-founder, Sun Yafang, each held 5 percent ownership of the company—and each ownership stake was by then valued at some $170 million.[64] Huawei was therefore considered a private company, or, in Chinese terms, 'siyou siying' (privately owned, privately managed).[65]

The company got off to a slow start in the late 1980s, as it served mainly as an agent for a Hong Kong company exporting switching systems and other equipment to mainland China. Ren soon realized he could make more money by manufacturing the equipment with his own company, and in 1990 sold switches for hotel networks at prices lower than comparable imported devices.[66]

As a private company, Huawei also faced the formidable task of raising capital. Unlike state-owned enterprises, which could rely on loans from the government-owned banking system, private companies in the early 1990s had few sources of funds. Huawei was forced to borrow from other large enterprises, at interest rates as high as 20–30 percent. But Ren and his cohorts invested wisely, spending as much as 100 million yuan (then about $20 million) on research and development. Huawei also established a research and development center in the US to get feedback on its products as well as technology updates.[67]

The company's investment in research and engineering personnel began to pay off in 1993, when it introduced its C&C08 central office switch. Ren had to use personal political connections, however, to get his first customer, a municipal office in the city of Yiwu in Zhejiang province. The successful deployment in the city led other Chinese state-run offices to consider Huawei's equipment. Huawei also had success in replacing equipment installed in the 1980s by Fujitsu and other Japanese companies in coastal provinces.

Despite the sizeable spending on research and development, which by the beginning of the next decade had reached 10 percent of the company's revenues, skeptics questioned the originality of Huawei's products. One critic familiar with the company claimed 'I have never seen anything that they make that is original technologically', and pointed out that the company could have reached its proficiency by reverse engineering rival companies' products, or by legally combing through existing patents to identify opportunities.[68]

In any event, a turning point for the company came in 1994, when Ren met China's Communist Party secretary general Jiang Zemin. In Ren's words: 'I said that switching equipment technology was related to national security, and that a nation that did not have its own switching equipment was like one that lacked its own military. Secretary Jiang replied: Well said.'[69] As noted above, in 1996, the government ended special import policies for telecommunications equipment, likely in reaction to national security concerns.

Huawei continued to expand sales, and, like ZTE, sold the majority of its equipment in the mid-1990s to rural customers. By 1996, Huawei had 20 percent of the Chinese switch market, second only to Shanghai Bell's share. The company's revenue that year reached 2.7 billion yuan ($325 million).[70] The next year, Texas Instruments (TI) cooperated with the company to establish a research lab, with TI supplying development tools and technical support.[71] In 1998, following the government's 'buy local' campaign, Huawei pulled even with Shanghai Bell, with a market share of about 22 percent.[72]

Huawei also used price wars to enter difficult markets. In 1999, for example, Huawei was ready to challenge Shanghai Bell directly in some of its own key markets. In Sichuan province, home to more than 80 million people, the Alcatel joint venture had 90 percent of the switching equipment market. Huawei offered their network entry product for nothing, but their competitors ignored the strategy. Following this early incentive, Huawei quickly increased its sales of switching equipment in the developing parts of the province, and came to rival Shanghai Bell's products. In this way, by 2002, Huawei had taken 70 percent of the market in the emerging areas of Sichuan.[73]

Like ZTE, Huawei also looked abroad, and was one of the few large private Chinese companies to establish an overseas presence. One of the company's first joint ventures was in Russia, where the company joined in 1997 with the Beto corporation to produce switching equipment.[74] By the end of 1999, the Russian partner was actively assembling imported kit equipment from the Chinese corporation.[75]

In addition to government support and the company's own marketing skills and spending on research and development, there was also a darker side to Huawei's method of doing business and gaining market share. According to a study by Cheng Dongsheng and Liu Lili, Huawei employed a complex system of agreements with local state-owned telephone companies that seemed to include illicit payments to the local telecommunications bureau employees.[76] During the late 1990s, the company created several joint ventures with their state-owned telecommunications company customers. By 1998, Huawei had signed agreements with municipal and provincial telephone bureaus to create Shanghai Huawei, Chengdu Huawei, Shenyang Huawei, Anhui Huawei, Sichuan Huawei, and other companies. The joint ventures were actually shell companies, and were a way to funnel money to local telecommunications employees so that Huawei could get deals to sell them equipment. In the case of Sichuan Huawei, for example, local partners could get 60–70 percent of their investment returned in the form of annual 'dividends'.[77]

By 2001, Huawei had begun transforming the shell companies into simple corporate branches; for example, Shanghai Huawei had become the company's East China branch.[78] Huawei may have been trying to legitimize these companies in preparation for a public stock listing, though a Huawei official noted in mid-2004 that the company had no plans to offer shares in the immediate future.[79]

At the beginning of the new millennium, the surviving joint ventures and Chinese companies faced new challenges to further building the telecommunications sector. Again, government policy shaped the way each player would contribute to the growth of the telecommunications resources.

Developments in the 2000s: Mobile and Internet Communications Equipment

By the beginning of the new decade, prices and profits for sale of fixed-line switching equipment were far below the earnings at the beginning of the 1990s, and equipment manufacturers began to look at the next wave of technology that could revive their revenues. As Figure 5.3 indicates, the next wave would prove to be mobile phone and Internet use, technologies that were becoming widespread among the Chinese population, in particular in urban parts of the country. Equipment manufacturing trends in these areas followed a similar path to that of fixed-line communications, and many of the same corporations took part in the growth of these new technologies.

Even before the beginning of the decade, the Chinese government put forward goals for market share of key parts of the growing mobile phone sector. In November 1998, the State Council and MII each issued reports calling for greater Chinese domestic corporate roles. The MII report specifically indicated domestic producers should begin to take market share away from the foreign joint venture manufacturers. For example, by 2003 the government's goal was for domestic companies to hold 70 percent of the market for mobile switches, 50 percent for mobile base stations, and 30 percent for handsets.[80]

Building the Network

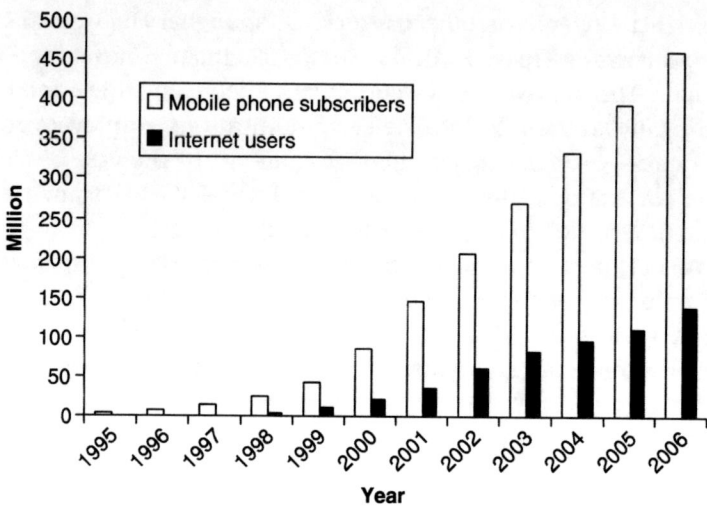

Fig. 5.3. Growth in number of mobile phone subscribers and Internet users, 1995–2006

Sources: China Internet Network Information Center (CNNIC), 'Statistical Survey Report on the Internet Development in China', Beijing (var. years); *Zhongguo tongji nianjian* (China Statistical Yearbook), Beijing: Zhongguo tongji chubanshe, 2004: 659; for 2004 and 2005 mobile phones: ibid. 2006, CD-ROM ver., chart 16–40; for 2006 mobile phones: MII website, at www.mii. gov.cn/art/2007/01/22/art_166_28236.html.

In a major difference with the earlier switching equipment pattern, China did not have to rely on imported mobile phone equipment in the early years of the network's growth—the joint venture companies could quickly move to provide necessary products. As we will see below, however, the country was initially dependent on imported equipment to build its Internet network grid.

Shanghai Bell, which had seen prices for its switching equipment erode in the 1990s, was one of the first companies to invest in mobile phone equipment manufacturing. In 1995, the Chinese central government approved the joint venture company's bid to begin production, and its new facility in Pudong, the rapidly developing manufacturing district in eastern Shanghai, opened the following year.[81] The company put a high priority on raising its level of technology, and put some 400 million yuan (about $48 million) into technology development in 1997.[82] It began selling equipment to build a mobile phone network for China Unicom in 2001, and the same year formed a joint venture with Korea's Samsung Mobile Communications Company to supply China's second largest mobile phone company.⁸³ By 2002, only about 50 percent of Shanghai Bell's revenue came from switching equipment sales; the rest came from selling mobile, terminal, and transmission equipment.⁸⁴

With reorganization of the MPT into the new MII ministry in 1998, the PTIC had to give up its stake in the venture. Still, the ministry maintained investment control with two MII-owned financial holding companies, named Hua Xin and Chang An.⁸⁵ In this scheme, the MII continued to reap financial rewards from the venture, even as its role as a joint venture partner was modified.

The Chinese government continued to lend its support to the Sino–foreign enterprise, as the Construction Bank of China gave the company 1.1 billion yuan (about $130 million) in buyer's credits in 1998,⁸⁶ and China's CITIC Bank offered 800 million yuan ($96 million) in loans in mid-2001.⁸⁷ The MII minister Wu Jichuan, who had taken the MPT leadership post in 1993, also seemed supportive of the French company that had joined with his own ministry's PTIC. In 2002, an Alcatel official described Wu as a 'visionary person,' with a 'genuine interest in Shanghai Bell development'.⁸⁸

Rival joint venture BISC had continued to sell fixed-line switches, even in the face of domestic competition, into the beginning of the decade. By the middle of 2002, the company had installed a total of 40 million ports in the country.⁸⁹ Sales peaked that year at nearly 12 million ports (with a profit of 800 million yuan, or $96 million), before declining to 10 million in 2002 and less than 10 million in 2003.⁹⁰ Like Shanghai Bell, BISC also moved quickly to offer alternative products, such as mobile switches and base stations.

An important change for Shanghai Bell came in late 2001, when Alcatel moved to take greater control of the company. In October, the French company announced it would pay $312 million to take a controlling share in the company, and would rename the venture Alcatel Shanghai Bell. The payment would give the company the Belgian government's 8.35 percent stake, as well as 10 percent plus one share from Huaxin's ownership; Alcatel would have a total of 50 percent plus one share, for a controlling interest. According to Alcatel officials, the Chinese did not want to sell more of the company, but they did effectively give the French controlling interest.⁹¹ Furthermore, the new company would incorporate other Alcatel wholly owned corporations in other parts of China into the new enterprise.⁹²

Alcatel decided it had to have full control of its China enterprise before it would introduce its most advanced technology for mobile and data systems. With the new controlling interest, Alcatel felt greater confidence in quickly sharing and integrating its cutting-edge technology with the company. 'Historically, only wholly-owned [subsidiary] companies get this technology... and without a controlling interest, we would not share the technology,' according to an Alcatel representative.[93] The move was further facilitated as China entered the WTO in late 2001. Greater promises of intellectual property rights protection further strengthened the incentive for Alcatel to take a larger interest in the company.

Siemens quickly followed the same pattern as Alcatel, though it faced considerably more resistance than had the French. Like Alcatel, in 2003, Siemens wanted a controlling stake in BISC to protect its technology, but the Chinese partners resisted. Siemens responded by suspending shipments of its high-technology components from Germany, thereby leading to a sharp drop in sales.[94] After six months of the German pressure, the Chinese side relented, and Siemens bought an additional 22 percent of the company in early 2004 to give it a controlling 62 percent ownership.[95] The remaining partners in BISC (after some previous shifting of ownership in the late 1990s) were Beijing Enterprises Holding Ltd. (a municipal conglomerate), with 20 percent, and 18 percent with the venture's other three original partners. The Beijing Wire Communications Plant (renamed Beijing Cable and Wire Technology Company) held only 6.6 percent.[96] Siemens now had control of the company, and would presumably be more willing to deploy its most advanced technology at its Chinese manufacturer.

Meanwhile, on the purely domestic manufacturers' side, Huawei had also begun to look beyond telephone switches to provide equipment for the rapidly expanding Internet data communications network. The American company Cisco Systems had been a major presence in China since the mid-1990s, and soon took a sizeable share of the market for devices such as data switches and routers, machines which forward data packets between computer networks. In 1999, Cisco had 80 percent of the China market for routers, but a familiar pattern began to emerge: in late 1999, Huawei also began to produce routers.[97]

Huawei soon began to chip away at Cisco's market, and by 2002, the Chinese company had 12 percent of the router market, vs 69 percent for Cisco. Huawei had achieved this market share in part by pricing its routers as much as 40 percent below the cost of Cisco's.[98] Cisco, however, suspected Huawei had stolen its intellectual property, and in January 2003, filed a lawsuit alleging the Chinese company had unlawfully copied some of its software.[99] The two companies settled the dispute in July 2004, with Huawei agreeing to revise some of its software. By that time, however,

Huawei's market share of routers and local area network equipment had risen to 31 percent, and Cisco's share had fallen to 56 percent.[100]

By 2004, the domestic telecommunications companies had begun to dominate much of the telecommunications equipment market in China. Huawei had become the big winner, with sales that year of $5.58 billion, and overseas sales at $2.28 billion.[101] It also had a 76 percent share of telecommunications equipment for the whole of China's domestic market.[102] ZTE had also seen rapid growth, with revenue of $2.7 billion in 2004, and overseas sales of $2.14 billion. ZTE continued to concentrate on lower-end markets in both China and abroad, with a focus on nations in areas such as the Middle East and Africa. It also became a major supplier of equipment to build China's local mobile phone network (known as *xiaolingtong* or 'Little Smart', discussed in Chapter 3) that grew quickly in the first years of the decade.[103]

For high-end products, however, Alcatel and Siemens would likely continue to strive for market share, using the ventures now under their management control and protection. In 2003, Alcatel spent some $1.8 billion on research and development, and Siemens about $2.2 billion, while Huawei spent only $385 million. Still, Huawei had some 6,500 patent applications as of the end of 2004, and, despite the accusations of reverse engineering noted above, seemed to be using its R&D money to good effect.[104]

What of the early MPT champion, PTIC, the company that seemed poised to be the dominant state player in the telecommunications equipment sector? As the MPT converted into the MII in 1998, PTIC was transformed into a government-owned holding company, and was renamed 'Putian' in 1999. As of 2005, it had continued to hold stakes in many of the same companies previously under its wing, with ten wholly owned companies, sixteen manufacturers, and eighteen partly owned companies. With revenue in 2003 of $7.2 billion, Putian continued to be the largest telecommunications equipment maker in China.[105]

Putian, however, had come to focus on mobile telephone handsets as its main growth engine. In 2003, Putian's companies (which included popular Chinese brands such as Bird, Eastcom, and Capitel) had about one-third of the national market, and other domestic producers took an additional 10 percent.[106] Putian alone had thereby met the MII's 1998 goal of Chinese domestic company control of the handset market.

It goes beyond the range of this chapter to analyze the dynamics of handset sales in China, but by mid-2005, foreign companies were riding a wave of Chinese consumer sentiment for state-of-the-art phones with high-end features such as megapixel cameras and sharp color screens. Even as Chinese companies cut prices, the market share for domestic producers was falling to as low as 20 percent, while foreign joint-venture manufacturers such as Motorola and Nokia, and foreign makers such as Samsung and others, saw big gains.[107] But as both ZTE and Huawei began to accelerate their own production of handsets in the middle of the decade, the market-share pendulum seemed poised to swing back to the Chinese domestic producer side.

Conclusion

The telecommunications equipment sector in China saw extremely rapid growth over the course of two decades. As previous chapters have indicated, by the early years of the new millennium, the country ranked first in the world in numbers of both fixed-line and mobile telephones, and second only to the United States in the total number of Internet users.

With the government playing a key function at nearly every stage of development, it is clear that the state played a significant and positive role in shaping the important industrial sector. Political leaders shifted neatly from a policy of allowing in imported equipment to one of cooperating with joint-venture manufacturers, and finally to fostering domestic state-owned and even private companies to take leading roles in building the communications network. The government set goals, used public money for financial incentives, allocated regulatory authority to the bureaucracy (chiefly in the form of the MPT and successor MII), and, in the latter years, fostered companies that could expand into international markets.

After a brief period of importing equipment, the government also used import duties to protect domestic companies from foreign competition. But it kept the door open to foreign participation in the form of successful joint ventures that competed with each other to offer China high-quality products at affordable prices. More significantly, the government orchestrated market competition among a variety of state-owned, foreign-invested, and private corporations. The result for the telecommunications sector was the building of a world-class communications network by foreign and domestic companies, and the growth of domestic Chinese companies that found strength both at home and abroad.

6

Telecommunications in Shanghai: A Case Focus on the Municipal Government's Role*

As Chapters 2 and 3 discussed, developed and developing countries have followed similar patterns as their telecommunications networks grew. In general, national policies, rather than municipal plans, have been dominant for building telecommunications networks. The need to finance the spread of the communications systems has made a national effort, in particular for developing nations, the general pattern for shaping the network in both urban and rural areas.

This chapter examines the ways a key city in China, namely Shanghai, built both its telecommunications infrastructure and operations services. It therefore highlights the policy process at the municipal level, and indicates how city officials can work to develop this key sector. As noted below, other cities in China had similar bureaucratic hierarchies, so the discussion likely reflects developments in other urban parts of the nation.

Why focus on Shanghai? From the mid-1980s, the city emerged as the most dynamic urban center in China. Shanghai's annual GDP growth from 1984 to 2006 averaged more than 16 percent. Annual income reached nearly 23,000 yuan (about $2,900) in 2006, nearly double the national average for China's city dwellers of 12,700 yuan (about $1,600), and far higher than national rural incomes of only 3,600 yuan ($450).[1] The city was becoming China's leading financial center, as well as a major manufacturer of automobiles and consumer electronics.

---

* Portions of Ch. 6 were previously published as: Eric Harwit, 'Telecommunications and the Internet in Shanghai: Political and Economic Factors Shaping the Network in a Chinese City', *Urban Studies* 42 (2005): 1837–58. Reprinted by permission of *Urban Studies*.

Shanghai: A Case Focus

The development of the telecommunications sector in the city of some 18 million closely followed Shanghai's rapid strides forward. In 1985, there was only one phone for every seventy people in Shanghai; fifteen years later, more than 97 percent of the city's residents had a phone installed in their home.[2] By the beginning of 2007, the city had more than 16 million mobile phones, some 5.1 million people were active Internet users, and the city boasted more than 150,000 .cn registered websites.[3]

Three sets of factors helped pave the way for the rapid progress of telecommunications in the city. First, government officials actively directed money toward building the network, and both national and municipal policies ensured Shanghai would receive necessary government support and, in some instances, strategic protection from foreign competition. This chapter describes some of the most important central and local government actions that contributed to the city's rapid telecommunications development.

A second element was the creation of municipal corporations that channeled funding into building both telephone and Internet data networks. The city government organized and supported companies that took leading roles in directing the funding, management, and expansion of key parts of the information infrastructure. As the chapter will show, however, by the beginning of the 2000s it appeared the city government's corporations had begun to take too large a role, and were beginning to stifle free market initiative.

A final part of the sector's growth in the city depended on the participation of foreign enterprises. Foreign equipment manufacturers arrived in the early 1980s, and foreign telecommunications operators made a tentative appearance in the mid-1990s. Both groups of investors made a significant mark on Shanghai's ability to develop its telecommunications industry, even as some of the government's policies limited their scope of activities.

After first putting the Shanghai case in an analytical context, the chapter briefly examines some of the city's historical technological development to assess its basic infrastructure as it entered its recent economic expansion stage. The chapter then turns to national and municipal government policies that directed funds for building a modern communications network. It focuses on the role of municipal corporations before considering how foreign companies, including Alcatel, AT&T, and others, also came to shape the city's telecommunications landscape. In assessing the growth of the Internet in Shanghai, it considers the case of the main municipal Internet content provider, Eastday.com, and how the city government's role shaped what Shanghai citizens see on their computer screens.

In its conclusion, the chapter evaluates the impact of central and municipal control in the sector. It also considers the effects China's entry to the WTO may have on the future course of telecommunications in the city. Finally, it assesses the potential utility of the Shanghai case as a model for cities in other developing countries.

Municipal Administration of Telecommunications

Municipal governments, lacking dedicated financial resources and the ability to compete with private companies or government monopolies, have typically played a relatively minor role in facilitating growth of network systems. In France, for example, local authorities in the 1980s tried to pressure the central government to create Zones de télécommunications avancées (ZTAs). These zones were to allow the municipal governments to use their own funds to subsidize construction of advanced telecommunications systems. France's 1982 'Plan Câble' gave responsibility for the country's TV cable network to municipalities. However, France Telecom was reluctant to allow the ZTAs to grow under local control, and the opening of cable systems to private competition in 1986 meant city governments were reluctant to spend public resources on building municipal networks.[4]

Still, several recent studies have explored the fundamental connection between telecommunications growth and a city's economic success, and indicate it is in the city's interest to have a well-developed communications network. Darrene Hackler's study of Minneapolis and Phoenix found a statistically positive correlation between the two cities' telecommunications capacity (as measured in Internet bandwidth) and the growth of urban high technology industry.[5] Paul Sommers and Daniel Carlson documented several cases in American cities in which large companies demanded developed telecommunications infrastructure before they would choose an urban area as a site for operations.[6] Susan Walcott and James Wheeler focused on Atlanta, and found high technology industry centered on areas of the city in which fiber optic communications capacity was highest.[7] Stephen Graham found that London, England, was a 'prime beneficiary of [the UK's] shift to competition, attracting intense competition from all the main global telecommunications players'.[8]

In contrast to most other developed and developing nations, China has given local government officials significant though limited budgetary power and decision-making authority to develop key areas of municipal economies. Jean Oi used the term 'local state corporatism' in 1992 to describe how 'fiscal reforms have provided incentives for local government officials to pursue economic, especially industrial, development'.[9] She found that local Communist Party officials were 'at the helm of the drive for economic development'.[10] Tian Gang pointed out that from the early 1980s, in southern Chinese cities such as Shenzhen, *'local governments* took more responsibility for their profits and losses'.[11] In more recent studies of high-technology industries, Adam Segal and Eric Thun noted that the Beijing municipality had 'an institutional endowment that lent itself to the development of small, innovative firms', and that high-tech firms formed in the city spread branch offices in other parts of China.[12] My own studies of the automobile sector in China indicated that the Guangzhou and Shanghai city governments played important roles in developing vehicle factories within the municipal borders, as they coordinated foreign investment and provided direct financial and support.[13] The Chinese central government therefore recognizes the important impact municipal decisions can have on local economic development, and has encouraged cities to work independently to develop important industries.

For developing an affordable and innovative telecommunications system, the evidence indicates competition is key to rapid growth. Most global telecommunications networks have been the result of private companies acting at a national level. In the case of China, however, central-government-inspired competition, together with significant municipal power, have given local governments the opportunity to guide the construction of this key industry to the benefit of their own metropolitan citizens. The case of telecommunications in Shanghai illustrates this unique set of circumstances, and shows that government ownership and management, when coordinated between local and national levels, can result in an initially positive outcome. As with other developed and developing countries, though, the government role can come to be burdensome, and may require a greater role for private domestic or foreign firms to further the sector's progress.

This chapter therefore builds on earlier analysis, and focuses on China's local-level telecommunications policy and its functions in concert with national government roles. It adds a new dimension in exploring how the city government of Shanghai exerted its influence on both the construction and operations aspects of voice and data telecommunications, and how foreign companies responded to government policies with mixed success.

Growth of Telecommunications in Shanghai to the 1970s

As Chapter 2 discussed, Shanghai was one of the first cities in China to develop telecommunications services. In 1871, the Great Northern Telegraph Company of Denmark laid an underwater telegraph cable from Hong Kong to Shanghai, giving the country its first submerged international communications system.[14] The presence of foreign traders and administrators in Shanghai from the 1850s was key to the growth of the city's early communications links.

In 1907 local telephone service was introduced, and long-distance service followed in 1923. World War II hampered the industry's growth, but in 1949, after the Communist victory, Shanghai had about 85,000 phone lines installed for some 55,000 subscribers. Though the numbers were small, Shanghai had almost 30 percent of all telephone lines in the country. The city boasted the largest manual exchange in Asia, with a capacity of 6,000 lines.[15]

Early Communist Party policy under new leader Mao Zedong followed the model of the Soviet Union, and emphasized dominance of state-owned industries, ones mainly engaged in heavy industrial production. Virtually all foreign and private telecommunications corporations were either nationalized or forced out of Shanghai and the rest of the People's Republic. With the new ownership structure, and the ideologically inspired lack of material incentive for innovation or efficiency, telecommunications development was slow in the 1950s and 1960s. At the same time, the central government feared economic dominance by large urban areas, and redistributed revenues away from developed cites like Shanghai to inland areas.

As noted in previous chapters, China's Cultural Revolution turmoil of the late 1960s retarded industrial growth across the country. But by the early 1970s leaders recognized the value of communications tools such as satellite systems and their terrestrial counterparts. In 1972, Shanghai was the site of the nation's first satellite earth station. Following the death of the radical Mao in 1976, and the rise of the more market-oriented leader Deng Xiaoping two years later, China's new regime put reform and modernization of industry as a top priority, and Shanghai received its share of attention in the modernization drive.

Key National Strategies in the 1980s and 1990s

As Chapter 2 indicated, in the late 1970s and early 1980s central government leaders realized domestic communications infrastructure and equipment could not meet the needs of a growing economy. Reforming Chinese firms

Shanghai: A Case Focus

were hampered by a scarcity of telephones, and connections were plagued by static, crossed-line background voices, and dropped calls. The problem was highlighted by a nascent but growing crop of foreign business people, many of whom resided in large cities such as Shanghai, who complained of long waits to make contact with corporate offices in Beijing or other parts of China or in their home countries.[16]

Chapter 2 discussed some of the key programs of the 1980s to increase telecommunications penetration, such as the 'Three 90 percents' policy of 1982 and other edicts to channel funding to the sector. As Figure 6.1 shows, the industrial take-off of the sector in Shanghai coincides with the later years of these policies. The number of fixed-line telephone subscribers in the city grew nearly eightfold, from about 290,000 in 1988 to 2.2 million just after the termination of the policy in 1995. On a national scale, the rise was an even faster ninefold over the same period, from 3.6 million to 32.6 million subscribers.[17] This rate was nearly double the overall GDP growth for both the nation and Shanghai in these years.[18]

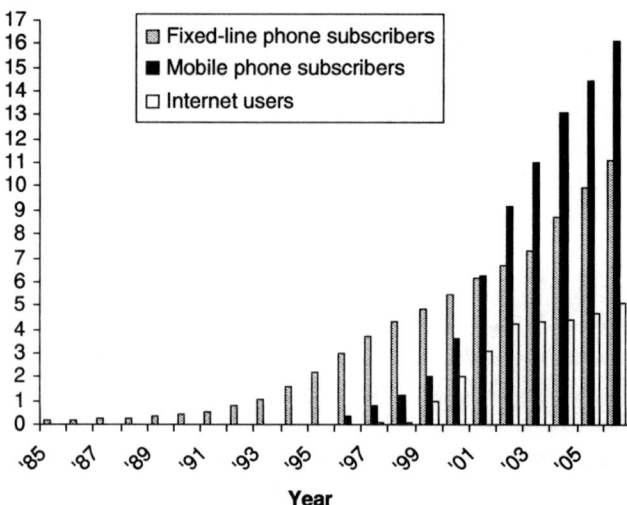

Fig. 6.1. Numbers of Shanghai fixed-line and mobile telephone subscribers and Internet users, 1985–2006

Sources: For 1985–2000 fixed-line numbers: *Statistical Yearbook of Shanghai*, Beijing: China Statistics Press, 2001: 212; for 2001–3 fixed-line numbers: *Shanghai Statistical Yearbook*, Beijing: China Statistics Press, 2004: 334. For mobile phone numbers, 1996–2000: *Statistical Yearbook of Shanghai*, 2001: 213. For 2001–3 mobile phone numbers: *Shanghai Statistical Yearbook*, 2004: 334. For 2004–6 fixed and mobile numbers: ibid. 2007: 296. For Internet numbers, China Internet Network Information Center (CNNIC), 'Statistical Survey Report on the Internet Development in China', Beijing: China Internet Network Information Center, var. year reports.

Shanghai: A Case Focus

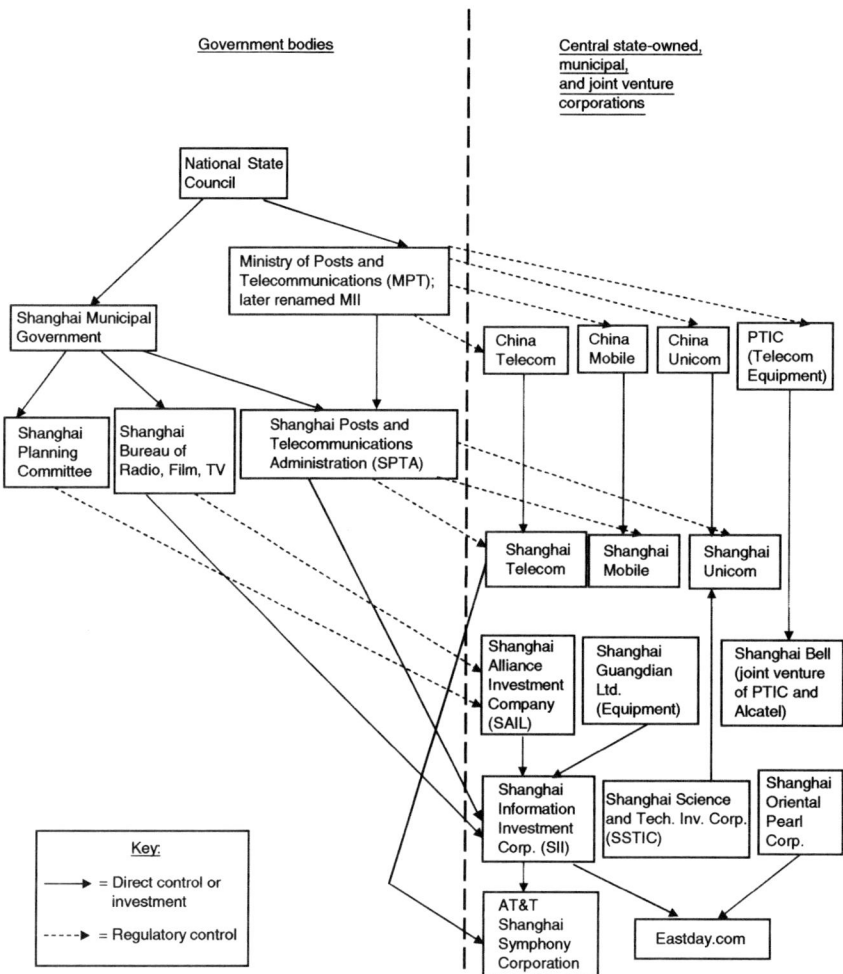

Fig. 6.2. Shanghai's national–local telecommunications hierarchy, with key corporations

The targeted telecommunications financial support was clearly helpful in facilitating the rapid expansion of telecommunications services at both the municipal and national levels.

Another major national policy that had direct benefits for Shanghai's telecommunications growth was the central government State Council's Directive 56. As noted in Chapter 5, this 1989 edict limited foreign participation in switching equipment manufacture to three companies: Alcatel of Belgium and France, Siemens of Germany, and NEC of Japan.

Shanghai: A Case Focus

Alcatel's main investment in China was its Shanghai Bell joint venture to make telecommunications switching equipment. Because Shanghai Bell's main partner was the MPT's PTIC, Alcatel's success had direct benefits for the nation's telecommunications regulator. Of course, it also had benefits for the Shanghai local economy.

Though the national government encouraged foreign investment in telecommunications equipment manufacture, as was the case with many other developing countries, it prohibited foreign or even private Chinese companies from taking part in operating the domestic phone companies (though, as Chapter 5 discussed, private Chinese companies were allowed to manufacture telecommunications equipment). As noted in Chapters 2 and 3, the Chinese government feared domestic security and social stability could be compromised if foreign or private corporations took part in running the communications network, and the MPT was also keen to keep profits within the country. The main state-owned companies of the 1980s and 1990s, China Telecom, China Mobile, and China Unicom, as well as their respective municipal phone companies, were prohibited from forming cooperative ventures directly with foreign corporations. However, as the case of Shanghai Unicom will show later in this chapter, in the late 1980s there was a concerted municipal effort to bypass this national policy.

Figure 6.2 indicates the relation between the national and local governments, and shows the position of some of the key government-owned enterprises involved in building the telecommunications network in the city. The following two sections, on municipal roles and foreign investment, consider the ways this political hierarchy shaped the growth of the industrial sector.

Municipal Policy and Telecommunications Expansion Initiatives

City leaders recognized early on that telecommunications would be an important part of developing Shanghai's economy. Measures to assist the industrial sector included implementation of financial policies meant to channel funds to expand the network, and creation of special municipal corporations that would link the city's interests to those of telecommunications operations and equipment sales.

The main office for regulating communications service was the local branch of the MPT, the Shanghai Posts and Telecommunications Administration (SPTA), whose leader was chosen by the MPT in consultation with local political leaders. The SPTA was both the city telephone regulator and operator until 1996, when a new entity, Shanghai Telecom (under the new national

China Telecom corporation), was split off to become the city's fixed-line phone company (see Figure 6.2).

In 1985, as new Shanghai mayor (and future Chinese president) Jiang Zemin took office, the city began to focus on developing its phone service. To emphasize the importance of the sector, Shanghai also created a special municipal 'telecommunications leading group'. The group included the city mayor, a relevant vice mayor, the SPTA chief, and other officials to coordinate development efforts. The group initially helped plan infrastructure installation for telecom development, and worked with the Shanghai price office to set appropriate rates for local calls.[19] The creation of this group was significant, as it indicated telecommunications would take high priority in the city government's plans for revitalizing Shanghai's economy.

The local leaders were quick to organize financial tools to foster the sector's growth. The municipal government used its power to regulate phone rates, and initially kept prices high to generate funds for the system's expansion. In the late 1980s and early 1990s, the cost of connecting a phone into an office or home ranged up to 5,000 yuan (about $940 at the time). This was a high fee for most city residents, whose average annual income was only about 2,500 yuan ($470) in 1991.[20] However, under the national 'Three 90 percents' policy, the city could reinvest nearly all of its profits from these fees to increase the purchase of telecommunications infrastructure equipment.

In 1986, for example, the city placed an order for enough digital switching equipment from Japan's Fujitsu Corporation to accommodate 60,000 new phone lines. This was a significant jump when Shanghai had only about 200,000 phone subscribers (see Figure 6.1).[21] From 1984 to 1987, as municipal equipment purchases continued, the number of phone subscribers grew from about 160,000 to 240,000, an annual rate of increase of about 14 percent. But following more active government steps to expand the network, together with a generally stronger municipal economy, the annual rate of increase grew to about 33 percent from 1987 to 1996, when the number of Shanghai fixed-line phones topped 3 million. Alcatel's joint venture company, Shanghai Bell, became the first domestic manufacturer of digital switches, and was a key contributor to the sector's growth at both local and national levels.

The city also early on took steps to make the business environment friendlier to foreign investors. In mid-1986, the city government helped arrange for joint venture companies such as Shanghai Bell to trade the Chinese currency they earned from domestic sales for the foreign currency needed to import equipment. At the same time, the Shanghai municipal authorities also eliminated import tariffs on raw materials used by the city's

Shanghai: A Case Focus

joint ventures.[22] The following section has greater detail on the ways foreign investment contributed to the city's telecommunications growth.

In the late 1990s, officials realized they had to cut the phone-line connection fee to allow even more city residents access to telephone service. By 1998, the fee had fallen to about 2,500 yuan ($300), and in February 1999, it was cut further to 1,500 yuan ($180), with no extra charge for installation of a second line.[23] With an average per capita income in Shanghai of about 8800 yuan ($1100) in 1998, the connection fee had become affordable to a mass market.[24] On 1 July 2001, as the number of new subscribers began to shrink, the city eliminated the installation fee altogether on private phones.[25] The rise in fixed-line customers continued in the first years of the new decade, but as the city neared saturation levels, the pace of installing new phones slowed.

The spread of mobile phones was another factor affecting fixed-line installation, as the wireless technology came to substitute for stationary phones. City officials oversaw a new set of policies aimed at expanding mobile phone access. The municipal government set mobile phone connection fees in early 1999 at about 1,000 yuan ($120), though handset prices ranged from 1,000 to 10,000 yuan ($120–1,200).[26] The rise of mobile phone competition in the form of Shanghai Unicom (see Figure 6.2) created a rival to the monopoly Shanghai Mobile corporation (itself split off from Shanghai Telecom in 2000), and contributed to a price war to attract customers. This chapter explores the Shanghai Unicom case in the later discussion of foreign investment policy.

For mobile phones, market forces came to play a larger role than government policy in expanding subscriber numbers. Fierce competition between mobile phone makers across the nation began to drive down costs, and by mid-2004, the least expensive handsets cost some 500 yuan ($60). However, government policy continued to be a factor, as the central Ministry of Information Industry (the MII, successor body to the MPT) set mobile phone usage rates at an affordable level. Per minute charges for outgoing local or incoming calls were 0.40 yuan to 0.60 yuan (5 to 7 cents), though competitive promotions between the mobile phone companies meant actual rates were even lower.[27] As Figure 6.1 indicates, by the end of 2006, Shanghai had more than 16 million mobile phone users, or about one mobile phone for every 1.1 residents. The number of mobile phones outnumbered fixed-line phones by nearly 5 million.

Finally, Internet subscriber growth in Shanghai matched national trends. In 1997, the city hosted only some 50,000 users, but the number had soared to 5.1 million by the end of 2006.[28] Again, market forces played a key role in expanding access to the data network, but city policies, centered on municipal investment corporations, also shaped the way Internet infrastructure grew. The following section discusses these corporations, and a later focus on the municipal 'Eastday.com' website and corporation illustrates the way the city has tried to shape Internet access within its borders.

Shanghai's Municipal Telecommunications Investment Corporations

One of the key municipal efforts for speeding the growth of the telecommunications and Internet industries in the city was close cooperation between Shanghai leaders and municipally owned corporations. Under the project heading 'Shanghai Infoport', the city government in 1995 launched an effort to improve information technology infrastructure for telecommunications, the Internet, and cable networks. From 2000 to 2005, the city planned to spend 70 billion yuan (about $8.5 billion) to improve the sector.[29]

The city directed much of this funding to several city-owned corporations set up to invest in and manage the building of the telecommunications and Internet cable and transmission systems. Figure 6.2 notes the placement of some of the key companies in the administrative hierarchy.

Shanghai was not the only city in China to utilize these municipal corporations to foster development. Beijing, Guangzhou, and nearly every other large city in the country had such entities that targeted areas the city government felt essential to the city's growth; these companies fit with the idea that local state corporatism could inspire regional officials to develop key sectors. However, scholars such as Segal and Thun, Tian, and Shen Xiaobai have indicated that Shanghai's companies in general had larger capitalization and a greater business scope than those in cities such as Beijing, Guangzhou, Shenzhen, or other parts of the nation.[30]

Within Shanghai, the Shanghai Alliance Investment Company Ltd. (SAIL) took a leading role in organizing the municipal corporations. SAIL was run by Jiang Mianheng, the son of the Chinese president and communist party leader Jiang Zemin.[31] (As Chapter 4 discussed, Jiang Mianheng had another finger in the telecommunications pie through his interest in the China Netcom data communications corporation). Though SAIL was formally an arm of the Shanghai city planning committee, Jiang Mianheng and his affiliates used the company to channel money to key municipal telecommunications projects. SAIL therefore was both a bridge to the central government, and a safe channel for directing funds to meet the Infoport project targets.

Shanghai: A Case Focus

The Shanghai Information Investment Corporation (SII), launched in 1997, was the main company designated to lead completion of the Infoport project goals. It was also backed by SAIL. As of late 2002, SAIL owned 37.4 per cent of the SII. The rest was held by the Shanghai PTA, the city's Bureau of Radio, Film, and Television, and its affiliated electronics manufacturer, Shanghai Guangdian Ltd.[32] SII's main projects included building the city's key high-speed infrastructure for data and broadcast media: these were the Shanghai Information Network, the Shanghai Cable Network, and the Broadband Fiber Network.[33]

The creation of the Shanghai Cable Network (SCN) in particular shows the way the SII received special consideration from the central government. Though the company's unique telephone voice service via cable was still not operating in 2004, the State Council had approved the project as a pilot for Shanghai. The company had already enrolled some 15,000 customers as of late 2001, and projected 80,000 by the end of 2002.[34] SCN did have more concrete success in offering 1 megabit per second broadband Internet access on its network, announcing in late 2003 a service plan costing slightly less than 100 yuan ($12) per month.[35]

As of the middle of the decade, Shanghai was the only city in China that was officially allowed to offer cable television, Internet access, and telephone services (a so-called 'triple play') through cable network lines—it had a special exemption from the rules noted in Chapter 4. Other cities had to keep telephone services separate from cable access. But, as the earlier chapter indicated, many other cities offered Internet access through cable television despite the opposition of the MII.

Other key SII investments, discussed later in this chapter, included one in a major Sino–foreign joint venture with the American company AT&T. SII had another hand in development of the Internet in the city, as it took a stake in a major Internet content provider, Eastday.com—this company is also examined below.

One other key municipal corporation involved in building the city's telecommunications infrastructure was the Shanghai Science and Technology Investment Corporation (SSTIC). The corporation was established in 1992 to be one of the city government's main venture capital investment arms, and specialized in high-technology fields including information technology.[36] As the chapter discusses later, the SSTIC, led by former vice mayor Liu Zhenyuan, was an important part of the effort to diversify the city's provision of mobile telephone services through the creation of the Shanghai Unicom corporation.

By the end of the 1990s, the city's telecommunications sector as a whole had become Shanghai's second-largest industry. Sales of wire and wireless communications equipment, terminals, and other related devices totaled 18 billion yuan (about $2.2 billion) in 1998. New city government policies announced in 1999 included better coordination in buying telecommunication equipment, with priority to local products; further investment and tax breaks to telecommunications companies; encouragement of foreign involvement with local companies; and guidance for technological development centers to integrate with manufacturers.[37] In early 2000, the official national *People's Daily* newspaper called the information sector in Shanghai the city's 'primary backbone industry'.[38]

The combination of national industrial policy to encourage the growth of telecommunications, along with active city policy and the encouragement of vital corporate transformation was a key to the rapid progress of Shanghai's telecommunications industry. In addition to national and local policy, however, foreign participation as well as local competition to the previous monopoly SPTA office also was an important part of the city's telecom development.

The Shanghai municipal corporations also seemed to avoid the stigma of corruption, even though Chinese development has been marked by state rent-seeking behavior since even before the Communist era.[39] However, the high-profile nature of Shanghai's endeavors may have helped insulate the companies' operations from problems of corruption; as Lü Xiaobo has pointed out, at the higher levels of the Chinese bureaucracy, supervision and control are tight, and officials have less discretion to perform illicit acts.[40]

Chapter 5 focused on one of the key telecommunications corporations in Shanghai, namely Shanghai Bell, and outlined the ways the central government and, to a somewhat lesser extent, municipal officials helped the company thrive in its first two decades. The following sections use studies of three other companies to focus on the ways the national and city governments worked to expand telecommunications infrastructure and service in Shanghai. In each of the cases, both central and local policies were integral to the companies' development paths.

Foreign Investment in Shanghai's Telecommunications: The Roles of Shanghai Unicom and AT&T

Shanghai Unicom

As noted above, foreign telecommunications companies have been officially prohibited from offering services in China. However, the former monopoly

Shanghai: A Case Focus

carrier China Telecom was not immune from a domestic challenge. Chapter 3 chronicled the development of the new national mobile phone company, China Unicom. One of Unicom's largest operations was its branch in Shanghai (see Figure 6.2).[41]

Shanghai Unicom was established in September 1994, and service was introduced in July 1995. Unicom also launched its first mobile phone service that month in Beijing, Tianjin, and Guangzhou, so Shanghai was not the only city in the nation to see a new municipal rival to the monopoly carrier. However, Shanghai was unique in that Unicom managed to establish its mobile service there earlier than the local MPT administration. The company signed up an estimated 20,000 subscribers before the Shanghai PTA launched its own mobile phone service.

Shanghai Unicom, like other city branches of the national Unicom company, was initially allowed to open itself to foreign company participation. In a move to circumvent the ban on direct foreign operations of telecommunications services, the Shanghai branch looked to form a so-called 'CCF' (Chinese–Chinese–Foreign) venture.

In late 1995, Shanghai Unicom teamed with the American company McCaw International, a subsidiary of Nextel Communications, to help further build the city's mobile phone project. Nextel pledged to invest some $22 million in the venture.[42] McCaw's partner in the 'CCF' arrangement, named Shanghai McCaw, was the previously mentioned Shanghai Science and Technology Investment Corporation (SSTIC). By 1996, as Chapter 3 noted, Shanghai Unicom was one of the few branch companies that turned a profit.

Through the municipal SSTIC, the Shanghai government took a direct stake in the new company's future. The connection was symbolized in the person of Liu Zhenyuan, a Shanghai vice mayor from 1983 to 1993 who subsequently served as both board chairman of SSTIC as well as director of Shanghai Unicom.[43]

The company was further reorganized in 1997, when the Hong Kong corporation CCT Telecom Holdings invested $22 million to take a 51 percent stake in the venture. The entity was renamed Shanghai CCT–McCaw Telecommunications System. McCaw's stake fell from 60 percent to 30 percent, and SSTIC held the remaining 19 percent. In turn, Shanghai CCT-McCaw held 60 percent of the Shanghai Unicom network, and the national China Unicom had 40 percent. By this time, Shanghai Unicom had nearly 50,000 subscribers.[44]

Though the company saw early success, Shanghai Unicom faced stiff competition and regulatory roadblocks from its competitor, the MPT's

Shanghai PTA office. The MPT matched and undercut Unicom's mobile handset prices and phone call rates, and made connection to the fixed network a large hurdle. For example, in late 1997, Shanghai Unicom charged 40 fen (about 5 US cents) per minute of incoming or outgoing calls, and gave the MPT only about 8 percent of this revenue; but the MPT kept 92 percent of Unicom's long-distance revenue, and 100 percent of international call fees.[45] Shanghai Unicom director Liu had earlier complained that 'when the MPT plays soccer, it also acts as referee. Of course it always wins'.[46] Still, Shanghai consumers benefited from the price competition.

In March 1998, Shanghai Unicom (like the rest of China Unicom) was merged with the MPT and other telecommunications bodies to come under the new Ministry of Information Industry (MII). By the end of the year, as Chapter 3 discussed, all 'CCF' operations were under review, and in early 1999 the central government announced they were to be dissolved. The desire to keep control of telecommunications operations in domestic hands had won out over the experimental foreign participation in the sector. Unicom itself looked to list its shares on a foreign stock exchange, and wanted to clear itself of the foreign ownership before it took this action.

In February 2000, CCT Telecom reluctantly reached an agreement to end its investment in Shanghai Unicom. It reportedly received between 800 million and one billion yuan (about $96–120 million) in cash and share options for its stake.[47] It is not clear how much McCaw received for its share, though on a proportional basis it would have received some 300 to 400 million yuan ($36–48 million). Considering both McCaw and CCT had initial investments of $22 million, the companies received reasonable returns under the circumstances.

Shanghai Unicom continued to accumulate mobile subscribers at a rapid pace, and by early 2001 had some 800,000 of Shanghai's 3.8 million mobile phone users.[48] By the middle of the year, Shanghai Unicom became Unicom's first branch to have more than 1 million subscribers.[49]

The case of Shanghai Unicom shows a further instance of the city government willing to host and assist foreign participation in its telecommunications sector, even when central authorities, in this case the MPT, were reluctant to support the practice. The city was able to absorb foreign technology, management skill, and capital to advance its telecommunications goals. Even though the foreign investors were forced out in the end, the company left in their wake continued to perform well and contribute to the sector's growth.

Shanghai: A Case Focus

AT&T

Though AT&T by 2007 was one of the major foreign telecommunications companies in Shanghai, it was relatively late in coming to the Chinese market. China invited the American company to examine joint switch-manufacturing prospects as early as 1979, but AT&T declined the offer. The company apparently was reluctant to transfer what was then seen as 'military-critical' technology to the PRC; it also was skeptical of China's production standards, and may have been harboring financial resources at the time for a domestic battle against American rival MCI.[50] Moreover, the 1989 State Council Directive 56, noted above, effectively shut AT&T out of cooperative switching equipment manufacturing efforts.

In the 1980s and early 1990s, the American company tried, though with little success, to manufacture other kinds of equipment in the PRC, and started its efforts in Shanghai. It used a Dutch subsidiary to create AT&T of Shanghai in May 1989. The company was owned 50 percent by AT&T's subsidiary, and the rest by the Shanghai city government's Optical Fiber Telecommunications Engineering Corporation (SOFTEC) and the MPT's subsidiary Shanghai Telecommunications Equipment Factory. The venture was created to make digital transmission equipment.[51] In 1993, a new AT&T venture, to manufacture optic fiber, was formed with SOFTEC, the Shanghai PTA, and the MPT's manufacturing arm PTIC (see Figure 6.2).[52] In 1996, following the split of AT&T, Lucent Technologies inherited the two Shanghai companies and four other AT&T joint ventures in other Chinese cities. But in late 2002, a weakened Lucent began selling off its China investments, with Corning International buying the Shanghai fiber optics company.[53]

After shedding its equipment-manufacturing arm, AT&T's core business was operating telecommunications networks to provide local and long-distance voice and data services. The new AT&T did not give up on its goal of expanding into the China market, though its main problem was the nation's ban on foreign participation in operating telephone services. AT&T did not choose, correctly as it turned out, to form a CCF agreement with Unicom.

Still, AT&T continued through the 1990s to discuss cooperation on operating a network. Talks began in 1993, with AT&T seeking a niche to help foreign-invested companies in China operate their communications networks, though the Americans also sought to provide services to a broader Chinese market.[54] The rapidly growing Pudong region of Shanghai was an early target area, and in 1994 AT&T agreed with the MPT and the Shanghai government to create an experimental high-speed network for the city. MPT minister Wu Jichuan, however, continued to oppose full foreign operations of telecommunications systems, even though the city government, with backing of then-mayor Xu Kuangdi, supported the company's goal.

In early 1999, the company signed a framework agreement to provide Internet-based telephone services in Shanghai, though operations were limited to Pudong. This was the only city in the country that allowed such foreign participation in telecommunications operations, and Shanghai's municipal government support was likely key to national government approval. In light of China's move toward participation in the WTO, MPT minister Wu eventually dropped his opposition to the project. Part of the reason for the breakthrough was Shanghai Telecom's desire, as the Internet boom began, to absorb data transmission technology from the foreign company.

AT&T's partners in the new venture were Shanghai Telecom and the previously discussed Shanghai Information Investment Corporation (SII) (see Figure 6.2).[55] The official approval for the project did not come until December 2000, when the venture was dubbed the 'Shanghai Symphony Telecom Corporation'. The venture had registered capital of approximately $20 million, with AT&T taking 25 percent of the company, Shanghai Telecom 60 percent, and SII 15 percent. The company would lease fiber optic lines already in place, and provide applications such as data services, virtual private networks, and other customized network operations. However, the company would not be allowed to offer voice telephone services.[56]

The key feature of the venture was AT&T's contribution of management technique, while it used Chinese infrastructure in the Pudong district. But a sensitive issue was AT&T's potential to challenge its partner Shanghai Telecom's customer base. The American company's targeting of foreign companies as customers for its services would minimize overlap, but eventual discord could result if AT&T were to become too successful. Though Shanghai Telecom chief Cheng Xiyuan led Shanghai Symphony's board of directors, few of the company's employees had ties to the Chinese partner's company.[57]

By the middle of 2002, the joint venture had successfully signed up some twenty local customers for its services, including Intel, the Chinese shipping company Cosco, and the Japanese corporation Omron. The key to further success for AT&T, however, would be to extend services to other parts of the nation. The company aimed first to expand its purview into the older and more populated Puxi part of Shanghai. Important future target cities were Beijing, Guangzhou, and Shenzhen. For 2006, the company reported a 32 percent revenue growth over the previous year for its China operations.[58]

Still, by the end of 2006, the company remained confined to Pudong. Though the city government had been important in facilitating AT&T's limited entry to the Shanghai market, the local phone company, Shanghai Telecom, was reluctant to allow the joint venture's further expansion in the metropolis. Both it and central government telecommunications officials wanted to keep the majority of telephone and data traffic, both in Shanghai and other cities, under Chinese control. Even China's 2001 entry to the WTO, discussed in previous chapters, did not create an immediate opportunity for AT&T to expand its stake in either Shanghai or other cities in the country.

Shanghai Municipal Control of the Internet: The Case of Eastday.com

As the Internet boom of the late 1990s enveloped the world, in China dot.com companies proliferated. Unlike the telecommunications operating corporations, many of the nation's data-content-providing companies were in private Chinese company hands. Chapter 4 noted some of the nation's most popular national sites included the Sina, Sohu, and Netease websites.

Of course, government at both national and local levels wanted to keep a grip on information content that appeared on computer screens. As Chapter 4 discussed, regulations prohibited dissemination of anti-Communist statements and promotion of rival political organizations, and online forums were to practice self-regulated censorship. The government also prohibited vices such as online pornography and gambling.[59]

This kind of political control by the city government on media content had a history in Shanghai. Lynn White's analysis of Shanghai news sources in the print and broadcast media in the 1980s pointed out the conflicting desire for control between central and local government sources. His work indicated that Shanghai publishers and broadcasters were able to shape most of their own reports and stories, in particular for information centered on local issues, except for topics deemed to have acute political sensitivity.[60]

Despite the mainly politically based censorship, a lively set of websites developed in areas of entertainment, culture, local activity, and other non-sensitive areas. By early 2007, the PRC hosted nearly 2 million .cn registered websites, and Shanghai had some 150,000 of these.[61] Cable companies such as the Shanghai Cable Network, as noted above, offered high-speed broadband cable service for 100 yuan ($13) or less per month.

As early as the year 2000, the Shanghai city government came to see the Internet boom as both a way to spread local news and other information, and as a source of revenue. The city was already the main Internet content provider, as most citizens used the local Shanghai Telecom phone company to gain modem access to the network at their home or office. But the government also wanted a share of Internet content revenue. While many other cities in China left Internet content to private providers who served both national and regional markets, Shanghai leaders decided to build their own municipal website to attract the city's netizens.

The new website was named Eastday.com. It was owned and managed by some of the key municipal corporations, including the Shanghai Information Investment Corporation (SII, the same Jiang Mianheng-backed company that partnered with AT&T); Shanghai Oriental Pearl (spun off in 1992 as a corporation from the city's Bureau of Radio and Television);[62] and major media outlets including print, television, and radio stations (see Figure 6.2). The company had registered capital of 600 million yuan (about $73 million). It planned to offer a range of services, including news and information, e-commerce, distance education, and entertainment services. Its goal was to compete against rivals by providing information before it appeared in other forms, such as print media or on-air broadcasts.[63]

The city opened Eastday.com on 28 May 2000, and the site almost immediately became the most popular in the city, with 11,000 registered users. It initially offered news, as well as articles on sports, culture, finance, and other mainly Shanghai-related information.[64] The website provided technology and networking platforms to some 150 media outlets in exchange for their news and information.[65] Other websites found it difficult to compete with the municipally backed Internet content provider.

As the Internet boom began to wane in late 2000 and 2001, the company turned to other means to generate profits. In October 2001, in the wake of a crackdown on illicit public Internet outlets in Shanghai (see Chapter 4), the company opened the first of a chain of Internet cafés in the city.[66] The company employed a franchise model: private businesspeople could open an Eastday-brand café, and pay Eastday 5 percent of the revenue they generated. In return, Eastday would help the café owner to obtain necessary city permits to begin and maintain operation. As a municipally owned entity, Eastday had a natural advantage for securing such permits. As noted in Chapter 4, independent private café owners were often subject to inspection and closure, so the clearly state-sanctioned Eastday presented them with a major rival.

By July 2002, there were some 238 Eastday cafés in Shanghai, with some franchise owners running up to five or six cafés. The cafés were relatively large scale, with the smallest having some 100 computers, and the largest about 800. Eastday cafés constituted about one-sixth of all of Shanghai's computer cafés, and about one-third of the total café terminals in the city. Customers paid 2 to 3 yuan (about 24–36 cents) per hour to use the computers, and many were young men playing interactive computer games.

With city backing and ease of registration, the cafés captured users from other, smaller, private Shanghai gaming parlors that had been periodic targets of government closures in the early part of the decade. As the Eastday franchises grew, the city's residents had more opportunities to find convenient access to the Internet. However, potential private competition stood to suffer as the municipal corporation dominated web access.

The Eastday cafés and the company's website therefore illustrate the city government's desire to tap financial returns from the rapidly growing interest in Internet use in the city. Both the websites and the cafés were potential tools to capture citizen spending related to web activities. The case also shows the city's strong hand in regulating Internet content and use. The popular website channeled official government information to the city's population, while the Internet cafés provided a tool for the government to monitor and potentially to censor information that café users accessed from their Internet sessions.

Though Internet use in Shanghai rose dramatically over its first decade (see Figure 6.1), the city government had made clear its intention to play a leading role in developing the medium's local content, and absorbing its online and Internet café revenue. Potential private rivals stood to find it difficult to compete against the municipally backed effort.

This case also indicates a divergence from other policies in other major cities in China. Beijing and Guangzhou, for example, had made far less effort to steer residents to municipally owned websites or Internet cafés. Shanghai's choice to keep a tight hand on physical access to the medium stood in future years to retard the menu of options for the city's growing body of netizens who lacked business or home access.

However, if the Eastday.com website failed to meet user expectations, the city's web viewers could easily choose alternative sites to satisfy their need for local news and information. Though the Shanghai model had given Eastday several important advantages, it was not clear the company would continue to dominate, even with government support.

The Course of Shanghai's Telecommunications Sector, and Lessons for Developing Cities

The overall course of telecommunications strategy in the city was one of guided encouragement and control, with municipal leaders and corporations keeping a hand in the main ventures meant to expand the sector. The chapter's case studies show both the national and municipal governments' caution in allowing either foreign corporations (AT&T, as well as Alcatel's Shanghai Bell) or even local companies (Unicom or potential Internet rivals to Eastday) too much competitive strength in Shanghai's economy. The results of government nurturing and protectionism were highly successful, giving Shanghai a strong base of infrastructure to further grow its economy.

Alternative paths, at least up to the end of the last century, may have been less fruitful. Had China or Shanghai allowed or encouraged private companies to build and operate telecommunications services and equipment manufacturing, the resultant companies may have had a difficult time finding capital from risk-averse domestic or foreign banks. State ownership guaranteed a source of funds to build public corporations quickly. In a similar way, opening the sector to free and open foreign competition in the early years of the industry could have brought domination by international corporations. Unlike other developing nations, like India and many countries in Africa, foreign equipment companies such as Alcatel, and operating companies including McCaw and AT&T, were allowed a share of the market, and brought valuable technologies that helped the Chinese build the framework for a modern and reliable communications network.

China's somewhat unique 'local state corporatism' gave Shanghai the power to channel even more resources into the critical area. As the research on cities in Europe and the US showed, a well-developed communications network was critical to attracting high-technology investment to the municipal area. Few other major cities outside of China had the ability or resources to help orchestrate the development of their telecommunications systems. Municipalities seeking to emulate Shanghai's path would need to secure cooperation from private or public telecommunications corporations, as well as from national telecommunications regulatory officials.

Shanghai: A Case Focus

Shanghai's development path faced tests in the years ahead, from potential foreign rivals as well as from domestic Chinese private industrial corporations. One of the greatest challenges for the telecommunications industry came with China's 2001 accession to the WTO. As noted in Chapter 3, under the rules of entry to the trade body China would allow greater foreign ownership of fixed-line and mobile phone ventures. It would also facilitate a foreign role in Internet and data services. The WTO also brought promises of greater intellectual property rights protection for foreign corporations. For Shanghai, this opened the potential for Sino-foreign joint ventures with core companies such as Shanghai Telecom, Shanghai Unicom, and Eastday, and many others, as well as greater potential property rights protection for companies like Alcatel.

To the middle of the first decade of the 2000s, Shanghai's municipal corporations played an important and positive role in building and operating the city's telecommunications infrastructure. However, their dominance over many parts of the industry could prove to be a repressive force on nascent private companies that could introduce greater and faster technological growth. The example of Shanghai Bell's fierce rival Huawei, which became one of China's largest and most successful private companies (as chronicled in Chapter 5), indicated that once a sector reached a sustainable level of growth, state-owned corporations should yield to the innovative power of the private sector. This step is a difficult one for a government that has built its legitimacy on public ownership of industry, but the trend in China from the 1980s was a march toward greater reliance on market forces. Though Shanghai's telecommunications sector seemed to lag this trend, city leaders could be forced to confront a greater private sector or foreign economic challenge in future years.[67]

Cities in other developing nations seeking to deepen their telecommunications infrastructure could look to the Shanghai case for ideas on how to harness and guide domestic and foreign capital and technology. The active participation of national and city governments could jump-start infrastructure building in cities where the private sector was unable to contribute sufficient capital. Government control over foreign investment would also ensure a developing nation's nascent networks would not become dependent on or fall under the control of a foreign corporation, one whose motives might have placed short-term profits over the rational growth of a city's information sector.

In the case of Shanghai, there was apparently little corruption in building the communications industry, except perhaps for the implicit power and prestige given President Jiang's son to orchestrate a key municipal corporation. As previously noted, the high-profile nature of Shanghai officials may have made corrupt actions more difficult, and the fruitful results of the government program in Shanghai indicate that if corruption existed it was not a major factor in slowing the sector's growth. Other nations' cities that may have looked to replicate the Shanghai example would be cautious that municipal officials refrain from taking personal profit alongside strong government involvement.[68]

In sum, the Shanghai case showed that a scrupulous municipal government, working with a supportive national government and foreign investors that remained, at least in the early stages, under some form of control, could bring about effective economic growth in a key sector such as telecommunications. If the national and city governments could later adjust their policies to include greater private business and foreign involvement once the sector had become more mature, growth would likely continue. Other developing nations that studied the Shanghai case could find a model that could help them develop their own networks in a rapid and effective way.

7

The Digital Divide of Telephones and the Internet*

In the framework of overcoming a global 'digital divide', China's development seemed to be a success story and perhaps a model for other developing nations. Previous chapters have chronicled the rapid expansion of telecommunications facilities. As of mid-2007, the nation had some 500 million mobile phone users, 370 million fixed-line connections, and 162 million Internet users. National teledensity stood at some 66 phones per 100 people, up from 0.2 phones per 100 in 1980. Multiple means of access to data networks, including high-speed broadband data transmission, was forming a key part of China's economic and social fabric.

Within China itself, however, the growth had left the nation's vast population with an internal digital divide among 'haves' and 'have nots'. For example, in the middle of the decade, the ratio of Internet access of the most connected parts of China, in the large cities such as Beijing and Shanghai, ran more than ten times that of the poorest regions of the country, such as Guizhou and Yunnan provinces. Telephone penetration rates ranged from near saturation in cosmopolitan cities like Shanghai, Beijing, and Shenzhen, to tens of thousands of rural villages without a single telephone. The Chinese government had begun to realize only slowly the potential problems associated with this internal urban/rural digital divide, as the disparity contributed to maintaining economic division among wealthy coastal residents and the hundreds of millions of inland countryside citizens.

In analyzing the PRC's digital divide, this chapter first puts the case of China in a comparative development context, and considers past studies of

* Portions of Ch. 7 were previously published as: Eric Harwit, 'Spreading Telecommunications to Developing Areas in China: Telephones, the Internet, and the Digital Divide', *China Quarterly* 180 (2004): 1010–30. Reprinted by permission of Cambridge University Press.

The Digital Divide

the importance of telecommunications spread to areas of low penetration. What are the potential benefits nations can receive by spreading communications access to a greater number of citizens? The chapter also considers how other societies may construct the idea of a digital divide, and formulates a working definition specific to the level of development in China. In building this definition, we begin to see the strategies and challenges for bringing equitable access to the population of a developing nation.

The chapter next turns to focus on the PRC, and examines the ways urban and rural communications systems have expanded over the past several decades. There has been open debate about the ways government can deal with issues of access to telecommunications, and the chapter considers some of the voices on each side of the policy question.

As the chapter will show, government leaders and institutions may have goals for developing the industry that may not necessarily take into account the broader needs of all sectors of society. Officials in charge of the telecommunications industry over the past two decades generally put revenue collection ahead of equitable access. Education officials, however, made some progress in spreading availability of data networks.

The chapter then considers the newer technology of data communication and transfer. In building on Chapter 4, it assesses problems related to equitable access to China's Internet. It then examines tentative government efforts to bridge the gap. The chapter concludes with an analysis of the effectiveness of these measures, and discussion of prospects of using technology to bridge the digital divide in the coming years.

The future course of the divide will likely be shaped by the interaction of these forces of technological change, institutional needs and capabilities, and equitable and fair access. The goals of government agencies may not conform with social desires and needs, nor with the ability of technology to provide services in cost-effective ways. The chapter and its conclusion consider how these factors will shape the future of China's telecommunications and digital data access.

Spreading Telecommunications to Developing Areas: Prior Studies and Context for the China Case

Several scholars have explored the importance of telecommunications access for developing parts of nations, and their work has relevance for the China case. For example, a study by Heather Hudson provided important evidence for the utility of spreading telecommunications tools to rural areas of both developed and developing countries.[1] Hudson's work chronicled a positive correlation between telephone spread and GDP growth, and indicated that two-way communications devices such as telephones (and, today, the Internet) were 'much more important factor[s] in the development process than one-way communications systems such as radio'.[2]

In the United States and other developed nations, empirical evidence indicates the utility of such expansion. As Chapter 2 chronicled, the American federal government took an active role in spreading communications to the countryside by providing loans to rural private telephone companies and to cooperative ventures. This support proved key to expanding access to communications systems, as the dominant private carrier, AT&T, was reluctant to offer services to less-lucrative rural areas. The result of this spending on rural telecommunications expansion, over the several decades from the 1930s to the 1970s, was an increase in GDP value of six to seven times the program's cost.[3]

In a study of mainly advanced European nations, Lars-Hendrick Röller and Leonard Waverman found a significant positive causal link between deployment of telecommunications and GDP growth.[4] The effect was more pronounced when teledensity rates reached 40 percent.[5]

In developing nations, Hudson cited a study of India indicating that rural residents derived economic surplus by using a telephone rather than traveling overland on their own to deliver a message in person.[6] In Sub-Saharan Africa, good telecommunications facilities contributed to greater ability to export agricultural products and raw materials.[7]

More recent evidence of the positive benefits of Internet connections in developing countries is less plentiful, though some studies indicate network access can be beneficial. One manager of the web café service Africa Online indicated that farmers in agricultural countries like Kenya were better able to manage their resources with access to information spread through the data network.[8] In Ghana, farmers used 'tele-centers' (combinations of phone and data network access stations) to find information enabling them to decide when to travel to a specific market to buy or sell products.[9] In general, it seems reasonable to assume that the benefits Hudson's study attributed to the two-way communication tool of telephones apply in the same way when citizens are given the ability to send and receive electronic mail, and to receive other data by using web browsers.

One of Hudson's other main arguments was that the peripheral economic benefits of telecommunications were often not factored into political decisions to extend access. The ability of telecommunications to generate revenues led to a profit–loss calculation that could leave poorer areas with little or no network connection. Hudson pointed out that governments often had little reservation to extend other infrastructure to developing regions—thus roads, water, and electricity were offered even when they might return little or no profit or even generate a loss, while telecommunications would lack similar subsidized support.[10]

The case of China, in particular during the era of a more market-oriented economy, mirrors this trend, as this chapter's evidence will indicate. However, the dilemma of Hudson's argument is that eventual economic growth in poor regions of a country may lead to positive revenue flows for telecommunications enterprises.[11] Chinese government efforts to bridge the divide could lead to quicker profitability for such rural-based telecommunications projects.

Defining the 'Divide' for China

The way the 'digital divide' is defined has obvious implications for narrowing the gap. How should the term be applied to a developing nation such as China?

Developing countries' leaders often think of the term as an international division of network access. Thus, in the case of China, former president Jiang Zemin commented in 2000 that 'the widening "digital divide" indicates that there exists a huge gap between the developed and developing countries in terms of the level of science and technology'.[12] In general, Chinese political leaders tend to think of the divide in terms of a gap between nations, though as this chapter will show later, there is an open debate in China about how to bring communications tools to those domestic citizens who lack them.

The Organization for Economic Cooperation and Development (OECD) has defined the divide rather broadly as 'the gap between individuals, households, businesses and geographic areas at different socio-economic levels with regard both to their opportunities to access information and communication technologies (ICTs) and to their use of the Internet for a wide variety of activities'.[13] Thus, we may consider the idea of a divide for developing countries to include basic access to telecommunications tools, as well as factors such as division along lines of regional residence, rural/urban location, and income level.

In the case of China, my definition of the divide accords 'communications access' to those with any sort of telephone, either fixed-line or wireless. 'Digital data access' is available to those with the rudimentary

The Digital Divide

tools for using data networks: a reliable supply of electricity, a wired or wireless connection, a computer or other device capable of accessing a data network, and minimal education (including basic literacy) needed for utilizing the technology. In this assessment, putting these tools in the hands of Chinese citizens is the method for bridging the divide.

In China, as this chapter will indicate, we can use comprehensive national data sets to find two major divisions of telecommunications and data network access. The first comes in the form of an urban–rural access divide, and consists of a contrast between generally affluent city dwellers and poorer countryside residents. The second is seen in a regional gap, between higher-income coastal Chinese, and those living in less-developed central and western regions. In both cases, income and educational levels play a role in the ability of citizens to have communications and network access. Chinese government statistics support the analysis used here, as numbers breaking down access along urban–rural and provincial lines are a standard method of recording economic data.[14]

Development of Telecommunications in Rural Areas, 1950s to the Twenty-first Century

Prior to the beginning of Communist rule in China in 1949, coastal cities with a large foreign-business presence had relatively advanced telephone systems. As the last chapter noted, for example, Shanghai had nearly 30 percent of all the telephone lines in the country, and its manual exchange was larger than even that of Tokyo.[15]

As Table 7.1 indicates, in the early 1950s most telephones were concentrated in the hands of urban dwellers. Some 86 percent of all phones in 1951 were in cities, yielding a six to one ratio of total urban to rural subscribers, and a penetration ratio (adjusted for population) of about forty-five to one. In the countryside, virtually all telephones were for official use; none was available to the general public.[16]

From 1953, the government began to turn its attention to the countryside with a campaign to provide phones at the county level. A 1955 Ministry of Posts and Telecommunications (MPT) conference set the more difficult goal of having every township connected to the national network by 1960. In the mid-1950s, the slogan '*xiang xiang tong dianhua*' ('telephones to every township') became a rallying cry as China's Great Leap Forward era of often irrationally planned growth began. A mass campaign in 1960 to expand irrigation networks in the countryside alone incorporated efforts to lay

The Digital Divide

Table 7.1. Number of fixed-line urban and rural subscribers, selected years, 1951–2006

Year	Total number of urban subscribers	Total number of rural subscribers	Phones per 100 urban residents	Phones per 100 rural residents	Ratio of urban to rural penetration
1951	273,600	45,500	0.41	0.0092	45
1955	375,200	103,800	0.45	0.020	23
1960	659,300	919,100	0.50	0.17	2.9
1965	771,104	492,222	0.59	0.083	7.1
1970	784,130	527,365	0.54	0.077	7.0
1975	1,032,827	659,220	0.64	0.086	7.4
1980	1,341,715	799,036	0.70	0.10	7.0
1985	2,189,554	930,744	0.87	0.12	7.3
1990	5,384,494	1,465,809	1.8	0.17	11
1995	32,635,600	8,070,000	9.3	0.94	10
2000	93,116,000	51,713,000	19.7	6.7	2.9
2002	135,791,000	78,431,000	27.0	10.0	2.7
2003	171,097,000	91,650,000	32.7	11.9	2.7
2004	210,251,000	101,505,000	38.7	13.4	2.9
2005	239,753,000	110,692,000	42.7	14.8	2.9
2006	251,329,000	116,456,000	43.6	15.8	2.8

Sources: For years to 1980: *Zhongguo tongji nianjian* (China Statistical Yearbook), Beijing: Zhongguo tongji chubanshe, 1981: 295; for 1985 and 1990: ibid. 1995: 495 and *China Posts and Telecommunications Annual Report*, 1996: 33; for 2000–2006: *Zhongguo Tongji Nianjian* (China Statistical Yearbook), 2007: 653. For rural and urban populations: ibid. 1993: 81, for population to 1990; *Zhongguo tongji nianjian* (China Statistical Yearbook), 2007:105, for population 1995–2006.

70,000 kilometers of telephone cable. The result of this particular effort was to add some 30,000 telephone subscribers.

According to government statistics noted in Table 7.1, by 1960, the country's rural areas had nearly 920,000 telephones, an increase of about twenty times over the 1951 number. Meanwhile, phones in urban areas grew only about 2.5 times. In fact, according to the numbers, there were now more telephones in the countryside than in the cities! However, analyses of Great Leap numbers by Western scholars assert many numbers from these years were inflated by local officials to show compliance with campaign goals, so it is not clear whether the large telephone increase of the late 1950s actually occurred.[17]

In any case, over the next five years, the actual number of phones in rural areas fell by more than 400,000, according to government numbers. During the Great Leap, sloppy installation of phone lines had apparently failed to follow an overall plan, leaving the network with many poor connections and overall low quality.[18] As a result, many of the lines (if they ever actually existed) were removed in the post-Leap years.

From 1962, the government turned its attention away from rural telecommunications development, and set its goal for network expansion again at the county, rather than township, level. The Cultural Revolution

The Digital Divide

of the late 1960s and early 1970s saw a modest expansion of the network, though poor maintenance practices meant network quality lagged.

From the beginning of his leadership of China's Communist Party in the 1920s, revolutionary leader Mao Zedong emphasized the importance of rural citizens to the success of the China's Communist system. However, the focus on providing phones and communications access to rural areas was likely more related to developing mechanisms of political control over the countryside through rapid communication means than it was an effort to broaden telephone access for farmers. For example, after 1956, the country's broadcasting and telecommunications authorities cooperated to use the phone lines for sending radio broadcasts to villages.[19] The vast number of the early telephone connections in both urban and rural settings were placed in public buildings, such as government offices or military command posts, rather than in citizens' homes.

From Table 7.1, we saw that the number of telephone subscribers in China's countryside during Mao's reign jumped from about 45,500 in 1951 to 659,000 in 1975, a nearly fifteenfold increase. Over the same period, urban phone lines, which still outnumbered rural connections, rose only about fourfold, from 274,000 to 1.03 million. Overall, as Table 7.1 indicated, the beginning of the Mao era saw a narrowing of the rural–urban penetration gap, from a ratio of forty-five in 1951 to about seven in the mid-1960s. However, there was little progress after the beginning of the Cultural Revolution to narrow the gap, which remained fairly constant to the mid-1980s. At that time, only about one in 100 urban residents had a telephone, and about one in 1,000 in the countryside.

The rise of Deng Xiaoping and his reform policies in the 1980s signaled marked change in China's economic system. As Chapter 2 discussed, telecommunications policy reflected the transition to a market-oriented society, and developments in the middle of the decade ignited rapid growth of telecommunications systems. The MPT could keep nearly all of its foreign revenue to further develop the network, many government loans for growing the system were forgiven, and provincial telecommunications offices were also allowed to plow their profits into further expansion. Rates for international call completion were set high to generate further income for the system's expansion, and foreign investment to build the system (though not operate it) was encouraged.[20] High connection fees of 2,000 to 3,000 yuan (about $240 to $360) per line in the mid-1990s also provided an important revenue source.[21] Government targets for numbers of new lines were rapidly met and exceeded by the late 1980s and early 1990s.

The Digital Divide

Over these years, the urban–rural telephone gap grew again rather quickly. As Table 7.1 indicated, in 1980 there were about 1.7 times as many telephones (seven times as many per capita) in cities as in the countryside. By 1995, attention to the city networks had led to an urban–rural ratio of four to one in total numbers of phones, and a per capita ratio of about ten to one.[22]

As with many other aspects of China's economic expansion, the reformist government had turned its immediate attention to infrastructure growth in the cities, and to a larger role for market forces in determining how investment would be made. Plans to open coastal cities as targets for foreign investment, as well as a concentration of industry in urban settings, dictated the need for rapid growth of telecommunications tools in these regions. It is therefore not surprising that the urban–rural telecommunications gap expanded during the first decade of China's new 'market socialism' economic system.

The expansion of telephone access into the home in the late 1980s and 1990s also allowed the relatively wealthier urban residents to accumulate greater communications access. The number of urban residential telephone subscribers rose from only about 41,000 in 1985 to 24 million in 1995; for rural residents, the increase was from 21,000 to 5.5 million.[23] The majority of telephones now were for residential use, but the increase was greater for those in cities who could afford them. Until the high connection fees were phased out in the late 1990s and early 2000s, the cost of installing a phone was one key barrier to residential phone service in poorer rural areas.

By the mid-1990s, after urban phone use had spread rapidly, the government again turned its attention to rural areas. The idea of spreading telephones to the countryside was revived, with the slogan '*cun cun tong dianhua*' ('telephones to every village') used to reinforce a commitment to provide telephones to even remote parts of the nation. In 1985, about 95 percent of China's 83,000 townships but only 45 percent of 940,000 villages had access to telephones.[24] In early 1996, MPT minister Wu Jichuan announced a goal of connecting each of the country's then approximately 740,000 villages to a telephone by the year 2000.[25]

The new policy had some slow though steady success. By 1998, 60 percent of villages had at least one phone.[26] A cut in the fixed-line connection fee from 2,000 yuan ($240) to between 500 and 1,500 yuan ($60 and $180) in March 1999 facilitated the more rapid spread of residential and business connections.[27] In late 1999, the number of connected villages had increased to more than 75 percent. The connection fee was eliminated in July 2001,[28] and in early 2003, 85 percent of villages had

The Digital Divide

telephones.[29] At the end of 2005, the goal had been nearly reached, as more than 94 percent of villages were connected.[30]

Table 7.1 showed that by the year 2000, the overall penetration ratio had narrowed significantly, at least for fixed-line phones, to an urban–rural ratio of about three to one, and the number of rural residential phones rose to nearly 46 million.[31] The gap continued to narrow in the first years of the new decade, as penetration rates in coastal and urban areas began to reach saturation levels. Still, residents of more than 100,000 villages had no access to a telephone within five kilometers of the village borders.[32]

Village access also varied widely by region. As Table 7.2 shows, many rural parts of wealthy coastal areas enjoyed near blanket phone coverage, while rural inland areas had many villages with no access. Geography contributed to the difficulty of connecting many of these regions, with

Table 7.2. Rural telephone access in selected areas of highest and lowest penetration, 2001 and 2005 (based on 2001 ranking of top 9 and bottom 5)

Province / Region	Percentage of villages with telephone access, 2001	Percentage of villages with telephone access, 2005
Beijing	100	100
Shanghai	100	100
Tianjin	100	100
Zhejiang	100	100
Jiangsu	100	100
Guangdong	100	100
Fujian	100	100
Henan	100	100[a]
Liaoning	94.5	100
Inner Mongolia	31	76[b]
Gansu	30.1	89
Qinghai	29.1	86
Guizhou	16.3	85
Tibet	4.3	35

[a] Henan province is listed as having only 96.2 percent of villages with telephones in 2004 (as noted on p. 172 in the 2004 edition of the *Zhongguo tongxin tongji niandu baogao* (China Telecommunications Statistics Annual Report), Beijing: Renmin youdian chubanshe. The 2001 figure is therefore likely somewhat inaccurate.

[b] Figure is for Inner Mongolia is for 2003. From *Zhongguo tongxin nianjian* (China Telecommunications Yearbook), Beijing: Zhongguo tongxin nianjian bianjibu, 2004: 374.

Notes: The report for the 2001 statistics categorizes 'access' as having a telephone available within five kilometers of the village border. The above source cites 'Annual Report from China Telecom in 2001' as its data reference, though does not indicate if the numbers are year-end 2000 or year-end 2001. For 2005, there is no indication where the phone is located related to the village; however, since the central government statistical source is likely the same, the definition also likely does not change.

Sources: For 2001: Research Center for Regulation and Competition, Chinese Academy of Social Sciences, 'Universal Service Obligations in China's Telecom Sector: Situations, Reforms, and Implementations'. (May 2002): 25. See this paper for a complete list of penetration rates by province / region. For 2005: *Zhongguo tongxin tongji niandu baogao* (China Telecommunications Statistics Annual Report), p. 210 for Gansu, p. 214 for Qinghai, p. 180 for Guizhou, p. 197 for Tibet; see notes above for Henan and Inner Mongolia. For developed cities and provinces, it is assumed that villages with telephones in 2001 retained them in 2005.

The Digital Divide

mountainous, hilly, and desert regions seeing special problems for providing connections. Of the unconnected villages, about 90 percent lay in mountainous or remote regions that were difficult to reach with fixed-line systems.[33] The concluding section of the chapter discusses various technologies that can overcome some of these geographical hurdles.

A rapid increase in the number of mobile phone users also has reflected a telecommunications divide. In 1993, China had only about 600,000 mobile phone subscribers,[34] but in early 2007 the number had increased to some half a billion. Wealthy coastal provinces and cities took the bulk of subscribers, with Beijing, Shanghai, Guangdong, Zhejiang, and Fujian in the top places in penetration, all at more than thirteen mobile phones per 100 people in early 2001. Guizhou, Tibet, and Gansu trailed at the bottom, with only about two mobile phones per 100 citizens.[35]

However, from the end of 1999 to early 2001, the penetration rate of the three lowest-ranking provinces had all increased by 100 percent or more, allowing them to keep pace with the top-ranking areas.[36] By the middle of the decade, many of the top cities and provinces had some fifty to more than 100 mobile phones per 100 people, but the poorest areas saw a corresponding rise: up to fifteen to twenty-five mobile phones per 100 citizens.[37] Table 7.2 also reflects the rapid spread of mobile phone services to more remote parts of the nation by 2005. As a later section of this chapter will discuss, the falling cost of mobile phone connections and handsets was a key factor in putting phones in the hands of many inland and rural Chinese residents.

Issues and Debates Over Addressing the Telecommunication Divide: Market Forces vs Government Intervention

One major problem for extending phone service in the countryside since the Mao years had been the government's ability to collect revenues. As market mechanisms became more important for regulating China's economy in the post-Mao era, bureaucrats' decisions on telecommunications policy came to conform more closely to the kind of profit–loss calculations noted in the above-cited work by Hudson.

From the 1950s, rural telecommunications losses had been common, and during the Great Leap years many party officials placed telephone calls for less than the official rate, or for nothing.[38] In 1979, the MPT proclaimed new rules for regulating the rural industry, and emphasized more rigorous collection of telephone fees. As a result, in 1981 rural telecommunications revenues began to exceed outlays, rising from a deficit of eight million yuan (about $5.3 million) in 1980 to a surplus of about six million yuan ($3.5 million) in 1981, and 91 million ($26 million) in 1986.[39]

However, by the mid-1990s, rapid growth in urban and coastal areas had made investment in these areas more profitable. The greatest generators of profit for the MPT in 1994, for example, were Shanghai, Guangdong, and Shandong, while the inland, rural provinces of Sichuan, Shaanxi, and Gansu all were net revenue losers.[40]

Revenue issues persisted into the following decade, as the phasing out of the connection fee made providing services to rural areas even less financially rewarding. In a poorer province such as Anhui, for example, by 2002 there was no fee to the customer for the installation of a first fixed telephone line. However, the actual cost of installation was about 2,000 yuan ($240), and telephone expenses income was only some 200 yuan ($24) per year.[41] Nationally, rural telephone service brought in a loss of 760 million yuan ($92 million) in the first 11 months of 2002, though this was an improvement over the 1.2 billion yuan ($145 million) loss in the same period of 2001.[42]

By the late 1990s and early 2000s, Chinese government officials and scholars began publishing opinions on how rural and inland development should best proceed, and focused on the issue of telecommunications access. For example, Xu Li, from the Ministry of Posts and Telecommunications' policy and legal regulation department, suggested in 1997 that it was most important to maintain growth in the developed coastal areas of the nation, and in particular build networks in the urban provincial capitals. Funds and technical expertise from the coastal parts of China should be used to develop inland areas, Xu argued, but the welfare of the whole network should not be sacrificed to develop poorer parts of the country.[43] The economists Liu Huaide and Hu Hanhui echoed this sentiment, acknowledging there was a gap between coastal and inland access to communications tools, but arguing that one should not artificially force the pace of telecommunications growth in poor areas. Liu and Hu asserted consumers in these areas may not see the need for or even understand how to use more advanced technologies such as Internet access.[44]

The prominent social scientist Hu Angang, long an advocate of equitable development, added his voice to the question of telecommunications access. In an article co-authored with Zhou Shaojie in 2001, he argued against relying on market forces to narrow the telecommunications and Internet gap. The government should play an active role by facilitating private and foreign investment in inland regions, and encouraging greater corporate competition for rural markets. The government should also mandate lower connection and service fees for developing regions, create a special fund to subsidize the spread of communications tools, and spur the convergence of technologies such as cable television with voice and data networks to provide greater avenues of access to less wealthy citizens.[45]

This debate illustrates the dynamic tension between bureaucratic interests intent on maximizing revenues and the political clout that accompanies such income, and advocates of equitable access as an entitlement. Up to the late 1990s, the government's preoccupation with economic growth in the advanced areas of the nation led to the triumph of the kind of policies advocated by telecommunications ministry officials. However, as we will see later in the chapter, the government took important steps in the early 2000s to achieve parts of the program advocated by Hu and Zhou. In particular, the introduction of new technologies, such as mobile phone equipment that could be deployed at a low cost, as well as less costly telephone and cable television connections for data services, stood to provide new tools not only for facilitating basic communications, but also for access to the data network of the Internet.

The Digital Divide of Internet Use

As in many developed nations, early use of the Internet in China came primarily from urban-based members of research and academic professions. Chapter 4 discussed the history of building the Internet, and indicated that most early users were scholars and students at the country's main universities and scientific research organizations.

Until the mid-1990s, then, much of the early Internet construction centered around major cities such as Beijing and Shanghai. In March 1995, the Chinese Academy of Sciences (CAS) linked branches in Shanghai, Hefei, Wuhan, and Nanjing as a first step to extending the Internet to the whole country. The CAS then turned to connect a further twenty-four major cities to the academic network, and the ChinaNET took similar action to build a national backbone network.[46]

Early commercial service was found almost exclusively in the wealthiest provinces, as Beijing and Shanghai municipalities, along with Guangdong, Liaoning, and Zhejiang provinces, began offering Internet access via ChinaNET's backbone in May and June of 1995. Most of this service was offered by regional telephone companies, though some private Internet service providers did emerge in these years. However, costs for usage were high, as

The Digital Divide

Chapter 4 indicated: monthly online fees averaged 400 to 600 yuan (about 48–72). These relatively high costs, at a time when average per capita monthly income even in urban Beijing stood at some 650 yuan (or $78) effectively limited use to wealthy city dwellers.[47]

As Chapter 4 also chronicled, in March 1999, at the same time it lowered telephone connection charges, the Chinese government moved to force down Internet access fees at several levels in order to make the service more affordable for private use. Hourly usage rates fell to average about four yuan (48 cents), or about 120–170 yuan ($14–20) per month.[48] By 2001, users could access the Internet from their homes using a modem, and have fees added directly to their phone bill. Costs were as low as 0.02 yuan (about 0.24 cents) per minute, and there was no need to subscribe to a separate specialized service provider.[49]

As Table 7.3 indicates, by the late 1990s the lower costs meant relatively large numbers of citizens in developed cities and provinces had access to the digital data of the Internet, while poorer, mainly inland provinces, lagged behind. Beijing emerged as the leader, with nearly 30 percent of its citizens classified as Internet users by the end of 2002. Shanghai and Tianjin also saw large numbers of users by the early years of the new decade.

At the low end, provinces such as Guizhou and Tibet had very few users, with less than 6 percent of each province's citizens getting access to the network. However, the annual rate of increase for Beijing users between

Table 7.3. Percentage of total population using the Internet in areas of high and low use, 2000–6, grouped according to top 5 and bottom 5 ranking areas in the 2006 count

	2000	2001	2002	2003	2004	2005	2006
Beijing	21	24	29	28	27.6	28.7	30.4
Shanghai	12	19	26	26.6	25.8	26.6	28.7
Tianjin	5.8	9.2	14	14.4	19.1	22.4	24.9
Guangdong	2.6	4.1	6.6	12.1	14.9	17.9	19.9
Zhejiang	3.3	4.8	7.2	9.7	11.4	15.0	19.9
Gansu	1.0	1.8	2.8	4.7	4.6	4.8	5.9
Tibet	0.3	1.2	2.3	3.2	2.6	3.3	5.8
Anhui	0.9	1.4	1.9	2.9	3.7	4.3	5.5
Henan	0.6	1.1	1.8	2.3	3.2	4.1	5.5
Guizhou	0.5	0.6	1.3	2.2	2.5	2.8	3.8
National average	1.8	2.7	4.8	6.1	7.2	8.4	10.4
Median	1.4	1.9	3.7	5.4	6.0	7.1	9.2

Note: According to the CNNIC, a user is defined as someone who has used the Internet for at least one hour per week, on average, for the past six months. User definition from summer, 2002 interview with a CNNIC official in Beijing.

Sources: Derived from China Internet Network Information Center (CNNIC), 'Statistical Survey Report on the Internet Development in China', Beijing: China Internet Network Information Center, var. year reports; population numbers from *Zhongguo tongji nianjian* (China Statistical Yearbook), Beijing: Zhongguo tongji chubanshe, var. years. For 2006 population: Xinhua Economic News Service, 1 Mar. 2007 report.

The Digital Divide

2000 and 2006 was only about 9 percent, while for Guizhou, the annual increase over the same period averaged some 41 percent and for Anhui and Gansu, about 35 percent.[50] The rate of increase among users in Beijing was nearly flat by the middle of the decade, while rapid growth continued in inland areas, indicating that the gap, at least on a regional basis, was narrowing into the latter part of the decade. Moreover, the median of Internet use was gradually approaching the average use percentage, indicating the development of more even distribution of Internet use across various provinces. But, overall, the growth of Internet use had begun to slow by the end of the decade, even in inland regions. This may have been due to users substituting mobile phone text messages for the popular e-mail function of Internet use.

Figure 7.1 graphically indicates Internet penetration across provinces. As expected, the relatively wealthier, urban coastal provinces show greater

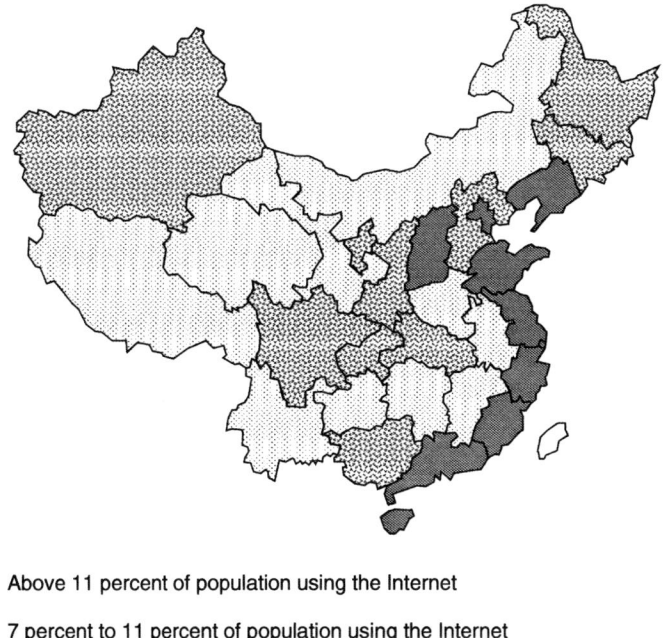

■ Above 11 percent of population using the Internet

▓ 7 percent to 11 percent of population using the Internet

□ Below 7 percent of population using the Internet

Fig. 7.1. Internet penetration, percentage by region, end 2006

Note: Map template courtesy of BDA (China) Ltd.

Source: China Internet Network Information Center (CNNIC), 'Statistical Survey Report on the Internet Development in China', Beijing: China Internet Network Information Center Jan. 2007 report.

The Digital Divide

Internet user levels, while many western provinces show lesser use. However, several provinces located in inland China, such as Xinjiang, Ningxia, and Guangxi have made progress reaching at least 7 percent penetration rates, and rates of growth in these provinces were higher in the mid-2000s than was the national average.

The regional growth shows a rough correlation with income levels. In 2006, for example, urban residents of both Beijing and Shanghai had average monthly incomes exceeding 1,800 yuan ($225), and rural residents of these municipalities had monthly incomes of more than 690 yuan ($86) (see Table 7.4). By contrast, urban Guizhou and Tibet dwellers had monthly incomes of less than 800 yuan ($100), and rural residents of these provinces earned less than 205 yuan ($26). With the cost of Internet use relatively uniform across the nation and starting at about 1 yuan ($0.12) per hour, it is not surprising the residents of poorer provinces showed less ability and interest in gaining Internet access.

Though most official government statistics on Chinese Internet use evaluated numbers along provincial and regional lines, the better way to assess user disparity may be along the larger national urban–rural split. In early 2003, China's Ministry of Science and Technology estimated that only 600,000 of the nation's nearly 60 million Internet users were living in rural areas.[51] This represented a ratio of about 100 to one. Among rural residents, who made up about 62 percent of China's population, this would mean a miniscule 0.08 percent of rural residents were Internet

Table 7.4. Average per capita monthly income for areas with high and low Internet use, in 2006

Region	Per Capita Monthly Income in Urban Areas	Per Capita Monthly Income in Rural Areas
Beijing	1868 Yuan / $ 234	690 Yuan / $ 86
Shanghai	1901 Yuan / $ 238	762 Yuan / $ 95
Tianjin	1290 Yuan / $ 161	519 Yuan / $ 65
Guangdong	1477 Yuan/ $ 185	423 Yuan / $ 53
Zhejiang	1663 Yuan / $ 208	611 Yuan / $ 76
Anhui	881 Yuan / $ 110	247 Yuan / $ 31
Jiangxi	835 Yuan / $ 104	288 Yuan / $ 36
Qinghai	817 Yuan / $ 102	197 Yuan / $ 25
Henan	862 Yuan / $ 108	272 Yuan / $ 34
Tibet	795 Yuan / $ 99	203 Yuan / $ 25
Guizhou	787 Yuan / $ 98	165 Yuan / $ 21

Note: Exchange rate for 2006 is one dollar equals approximately 8.0 yuan.
Sources: Derived from annual income statistics in *Zhongguo tongji nianjian* (China Statistical Yearbook), Beijing: Zhongguo tongji chubanshe, 2007: 355, 369.

The Digital Divide

users, while about 12 percent of the nation's urban residents used the network.[52]

Figure 7.2 gives a rough summary of the status of the information and communications technologies in the urban–rural divide. The progress to the middle part of the decade in fixed-line phone connections for the vast majority of Chinese citizens was significant. As the chart points out, however, the rural areas lagged behind developed parts of the country in more advanced communications tools. The prospects for the spread of higher-level technologies, such as dial-up Internet and broadband access, are discussed in the final section of this chapter.

What were some of the main problems for bridging the rural–urban and regional Internet divide, and what steps was the Chinese government taking to bridge it? First we examine basic problems for spreading telecommunications and Internet access to rural areas, and then consider ways the government could move to narrow the gap.

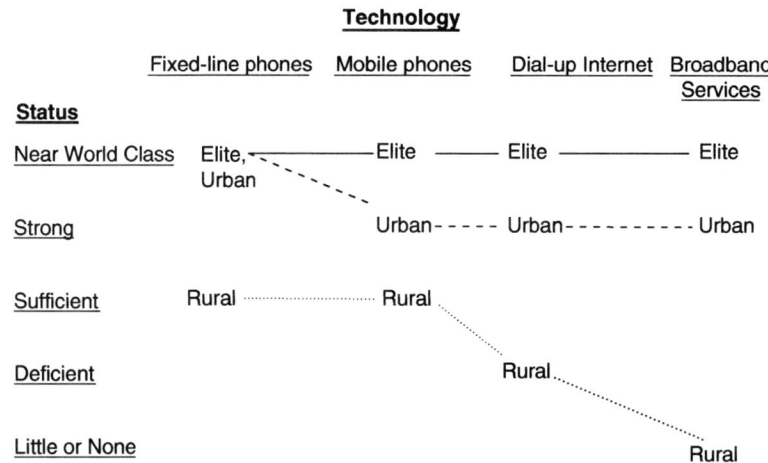

Fig. 7.2. Status of the telecommunications revolution in elite urban, urban, and rural areas

Notes: 'Status' is a subjective measure, taking into account factors such as availability, quality, and affordability of the technology. 'Elite' refers to citizens with a high level of disposable income in urban areas; 'Urban' represents average urban residents; and 'Rural' refers to any resident of countryside areas.

173

Problems of Spreading Digital Data Access to Rural Areas

There were several key problems in bridging Internet user disparity. Noted earlier in this chapter was that users needed both a reliable way to connect to the network, and equipment to display and manipulate the results of a connection. Users also needed the basic technological skills to utilize and interpret information from the network; basic literacy was naturally a key ingredient. On a more rudimentary level, a constant supply of electricity was also necessary to allow one to access the Internet. What were obstacles to overcoming these barriers?

The first part of this chapter focused on progress of spreading basic telephone service to rural and poor areas of the country. As some 39 million (about 28 percent) of China's Internet users in 2007 employed telephone lines to get access via a phone modem,[53] the lack of communications connections for many citizens also hindered access to the data network.

A further problem in giving data access to rural communities was the lack of basic infrastructure. For example, in 2002 some 30 million of China's citizens, living mainly in the northwest rural parts of the country, lacked access to electricity. A far larger number had poor or inadequate power supplies.[54]

Income level disparities exacerbated the problem. As of 2004, the annual disposable income gap between urban and rural residents was 3.23 to 1.[55] As noted earlier, inability of rural and inland citizens to pay for Internet use fees was a major obstacle to expanded access for less wealthy Chinese. And, of course, the cost of a computer was beyond the means of many rural residents.

Low per capita income in poorer provinces could also lead to lower rates of educational attainment. Rural children often dropped out of school to work in fields or local industry, and even minimal costs associated with schools could be prohibitive for the poorest farmers earning less than 1,000 yuan (about $120) per year.[56] Without basic educational skills, the value of Internet access was incomprehensible to many inland Chinese citizens.

Of course, basic literacy also limited citizen access to Internet information. As of 2006, some 9 percent of the country's population above 15 years of age were officially classified as illiterate. The rates were higher in the western provinces at nearly 17 percent in Yunnan, 19 percent in Guizhou, and 46 percent in Tibet, where Internet penetration was among the lowest in the nation.[57]

Education disparity at higher levels may also account for some of the differences in Internet penetration, even in regions of the nation that were generally poor. For example, as Figure 7.1 indicates, inland provinces such as Xinjiang, Guangxi, and Ningxia had higher Internet penetration rates than those of some of their neighbors. These provinces all had income levels that fell within the bottom tiers of both the urban and rural regions of the lowest-ranking areas.[58] However, among residents of Xinjiang, Guangxi, and Ningxia, as of the year 2006, from 16 to 20 percent had senior middle school education or higher, while in Tibet, Guizhou, and Anhui, only 4 to 14 percent of residents had an equivalent level of education.[59]

Less exposure by inland provinces to high technology and foreign influences also acted as an impediment to the spread of Internet use. A 2002 study of poor rural areas in Sichuan province indicated many users had no notion of the Internet. For them, even the idea of owning their own telephone was a distant dream.[60] Few rural residents in the regions had any access to computers, or to other devices capable of making connections to data networks.

One final issue for extending services to poor, rural areas may have been government hesitance to give discontented citizens access to tools that could allow organized opposition to a one-party government system. In the event China's larger economy faced significant hardship, as it did in some of the years preceding the 1989 Tiananmen student demonstrations, the Internet could indeed become a useful means of coordinating opposition voices.

The political ramifications of Internet access for disadvantaged and discontented residents also were key to understanding the ways telecommunications tools could spread, as noted in Chapter 4. Up to 2007, the government had put up few ideological barriers to obtaining basic Internet access. Any citizen with money could use the data network, as long as he or she avoided posting or transmitting information that would directly threaten government interests. There was no evidence of any special rules that would deny access to rural, inland, or poor residents based solely on their social or economic standing and potential use of the Internet for anti-government purposes.

Chapter 4 discussed several cases of Internet dissidence that arose in the first decade of the 2000s, ones resulting in the arrests of those posting information officials found critical of the government or promoting anti-government group formation. The government therefore faced some contradictory pressures, wanting to provide communications tools' economic, educational, and other gains, while mindful of the utility these technologies could have for those opposed to Communist Party rule. It was likely that the publicized arrests of those posting unacceptable information, along with a still significant police presence at many levels of

Chinese society, instilled a feeling of self-censorship among those tempted to use the voice and data networks to challenge the one-party state.

One danger to the government, however, could arise with organized use of mobile phone messages or perhaps e-mail to coordinate an anti-government effort in an area of focused discontent. Short-message services (SMS) allowed mobile phone users to send electronic notes to group members instantaneously, so that rebellious leaders could skillfully choreograph rallies or marches. These messages would be more difficult to monitor than, for example, publicly posted website information. However, the government's firm control over the communications backbone would likely allow it to shut down a public mobile network or e-mail service in an affected area, with the police retaining alternative means of communication. The government would still maintain control of telecommunications tools, even in a crisis, thereby diminishing the threat that these tools could be used by anti-government forces in the time of crisis.

Programs to Bridge the Urban–Rural Digital and Communications Divide

From the early 2000s, the Chinese government formulated programs to further both communications and digital access for inland, rural citizens, and new technologies that could also facilitate the bridging of the divides. This section examines attempts to spread access more equitably to Chinese citizens; the concluding section explores newer technologies that could further assist in narrowing the gap.

In the case of the communications divide, the earlier part of this chapter pointed out the policy priority the government had assigned to spreading basic telephone access. In the early 2000s, the government considered the formation of a Universal Service Fund (USF) along the lines employed in the United States and other developed nations.

In the American case, Chapter 2 already outlined steps taken by the federal Rural Electrification Administration and its successor agencies to fund expansion of communications tools to rural areas. This was supplemented by a USF, supported by taking a small portion of interstate long-distance revenue to reimburse local phone companies that incurred particularly high costs for connecting customers to the public network.[61] The Telecommunications Act of 1996 modified the USF, mandating that all telecommunications companies providing either fixed-line or mobile service had to contribute to the fund. The fund supported not only rural access but also gave subsidies for any low-income subscribers, rural health-care providers, and schools and libraries seeking to add communications and Internet access.[62] In 2006, the American USF distributed about $7.3 billion toward these services.[63]

China's Ministry of Information Industry started research on a universal service obligation as early as 1998, after the split of the ministry's telecommunications and postal operations. In May 2000, the State Development and Planning Commission (SDPC) suggested the establishment of a Universal Service Fund in accordance with international standards. The fund was to be used for both telecommunications network construction and infrastructure maintenance in rural areas, as well as for the construction of a private network for the Chinese Communist Party and the government.[64]

Steps to develop a USF were put on hold as the government moved to reorganize its basic fixed-line phone companies. As Chapter 3 discussed, in 2001, the government announced it would split the monopoly carrier China Telecom into firms serving the nation's northern and southern regions. In the north, China Netcom took over from the previous monopolist, China Telecom. The southern part of the nation was left under China Telecom's jurisdiction. The most needy beneficiary of a USF would be the new China Telecom because it inherited the poorest western provinces and lost the ability to compensate for losses there with income from many of the wealthy northern coastal provinces.

In August 2001, the MII sent a team to the United States to study how the American federal government calculated the cost and management mechanism for universal service. In June 2002, the MII submitted a study report to the Ministry of Finance on the feasibility of setting up a USF. In the second half of 2002, the two ministries planned to discuss the details of how to raise and manage the fund. The recommended size of the proposed fund was 10 billion yuan ($1.2 billion).[65]

Finally, in early 2003, the MII announced all of China's major fixed-line and wireless telephone companies would have to pay an annual fee of 0.24 yuan (about 3 cents) for every phone number they owned (even ones not currently in use) in order to spread phones to rural regions. The total projected revenue would be about 400 million yuan (about $48 million) per year.[66] This seemed considerably short of the necessary total fund size projected in the previous year. However, it was a meaningful beginning in the attempt to subsidize carriers who provided telephone service to regions that generated losses for the communications service providers. As of mid-2007, the USF was still in its planning stages, and had not been deployed.[67]

The Digital Divide

China also pursued expansion of digital data access on various fronts. For example, one of the leading forces in spreading data services to China's schools was the Ministry of Education's China Education and Research Network, or CERNET. The network was founded in 1993, and had as its goal the provision of Internet connections to universities as well as secondary and primary schools. CERNET did not build all of its own data network, but rather leased lines from the major telecommunications operating companies, such as China Telecom and China Netcom.

CERNET's first target was China's universities, and the organization had established Internet connections on about 100 campuses by 1996.[68] By mid-1998, the number had risen to 300 universities and some 200,000 users, and an average of 2,000 computers per campus.[69] The following five years saw continued growth, with nearly all of China's approximately 1,400 universities connected by CERNET in early 2003. At this time, the network claimed some 10 million users, and was also the only Internet service provider in Tibet.[70] As of late 2005, CERNET claimed more than 20 million users.[71]

CERNET's focus on tertiary education was primarily linked to the need for providing low-cost service. The company, subsidized by the Education Ministry, charged the institutions an annual average cost of only about 20 yuan ($2.40) per student. This compared with a monthly charge of some 120 yuan (about $15) for access to the commercial public ChinaNET service. In some cases, university connections did include restrictions on international network access because this would add to the service cost. However, basic functions such as e-mail and domestic access were included in the higher education service.[72]

Despite progress at the university level, by 2005 CERNET had connected only about 200 secondary schools and only a small number of primary schools, ones mainly in the wealthy coastal areas. This represented only a tiny fraction of the country's approximately 79,000 secondary schools and 390,000 primary schools.[73] The main limiting factor was cost, and CERNET, though apparently eager to continue its network expansion beyond college campuses, seemed to have stalled for lack of funding.

Other governmental and corporate efforts attempted to address the basic problem of computer access in schools. In early 2001, for example, China Great Wall Computer Group announced a project to invest 150 million yuan ($18 million) to supply 300 primary and high schools in inland Henan province with Internet-access instruction rooms.[74] However, much of the investment in computers for educational purposes still came in coastal regions. For example, the Beijing municipal government gave some 7,500 computers to the city's schools in 2001,[75] and Shandong province commissioned the Tengtu International Corporation to link 40,000 of its K–12 schools over the years 2001 to 2014.[76]

Other than these programs, interviews in China in mid-2002 indicated the government was taking few steps to invest in spreading communications access to Chinese 'have nots'. Even some of the steps noted above tended to favor educated, wealthy citizens, though some, such as provision of access to college students in poor regions, may have had more direct benefit to those in greatest need.

Rural residents in particular stood to lose out as the data network was slow to penetrate their towns and villages. New services such as distance medicine promised to provide higher levels of medical care to those unable to find access to specialized doctors in their home region. Distance education could be another tool to bring knowledge from advanced urban instructors to rural or poor areas of the country. A report from the International Bureau of Education at the United Nations Educational, Scientific, and Cultural Organization (UNESCO) indicated that 'the evidence is consistent that, given adequately large numbers, the cost of distance education is likely to compare favourably with those of conventional education...'.[77] China's lack of data and telecommunications connections would be a major impediment to the spread of such services.

Falling Prices, New Technologies, and the Future of the Digital Divide

A 2002 Chinese Academy of Social Sciences report indicated that wireless and satellite technologies were likely to be key tools available for reaching the approximately 15 percent of villages that, at the time, lacked phone access within five kilometers of village borders.[78] As noted above, of the unconnected villages, some 90 percent lay in mountainous or remote regions that could not be served by fixed cable systems. For about 5 percent of these villages, ones with scattered populations residing more than fifty kilometers from a telephone, only satellite technology could bring a connection.

By the middle of the decade, rising costs of fixed-line installation and falling fees for setting up mobile networks had made mobile systems up to 80 percent cheaper to install. A wireless system could cost only ten yuan (about $1.20) per month per line to operate, while the comparable cost for a fixed line was 30–40 yuan ($3.60–4.80). The cost of installing a satellite line could be more than ten times that of a mobile line, and maintenance

The Digital Divide

costs were also great.[79] Therefore, it was likely that much of the revenue from the proposed Universal Service Fund would be channeled to creating mobile (though perhaps not satellite) networks in villages with few or no communications services.

From Table 7.2, then, we can see marked progress by the middle of the decade in the spread of villages that had at least one telephone for local inhabitants. The falling costs of building mobile networks had facilitated much of this expansion. China Mobile, the major wireless provider, also stated it would 'guide handset vendors' to sell low-price mobile phones in rural areas—by 2006, prices of the phones even in large cities had fallen to 400 yuan (about $50) or less. The sale of low-cost phones would help the operating companies to expand their market in these regions. Even foreign phone manufacturers took part in the expansion to the less-developed parts of the country: Finland's Nokia indicated it would market handsets with symbols, rather than letters, for users who might be illiterate.[80]

As Chapter 4 discussed, another method for expanding data and even phone access could come with greater convergence opportunities for transmitting voice and data. For example, as of the end of 2006, China had some 138 million cable TV subscribers, who stood to enjoy more interactive services once regulatory issues were resolved.[81] Though there seemed to be no plans to subsidize high-speed broadband access for poor or inland rural regions, market forces stood to play a role in lowering user costs and expanding the customer base for this technology. Following the government-mandated creation of rival companies China Telecom and China Netcom in 2002, competition between the two companies' data networks helped raise the number of those using high-speed connections to the Internet. Monthly rates for 1 Mbps unlimited broadband access in Beijing fell to as little as 150 yuan (about $18) in early 2005.[82] Other areas saw rates as low as 70 yuan (about $8.50) per month.[83] However, even at this cost many of the users would likely be wealthy city dwellers.

Another new technology that in theory could greatly and quickly ease the digital divide was the transmission of Internet data information as well as voice traffic through ordinary electrical power lines. Though still in the experimental stage as of 2005, commercial use of power lines for Internet access had become available in some parts of the United States. Users paying $30 per month could get access at four times the speed of an ordinary telephone modem.[84] As China extended the very basic necessity of electricity to rural areas, the ability to use ordinary power sockets as data outlets opened a new dimension to rural communications access. Having one cable provide electricity, telephone, and Internet access would solve the key issue of providing a data pipeline, though other problems, such as access to computers, training, and basic literacy, would remain in many areas.

The tension and balance between the forces of technological change, bureaucratic interests, and equitable access had begun to shift in the new millennium. Though the Mao era saw equity as a primary goal, limits on technology and political will and ability hampered efforts to spread telephones in any meaningful and lasting way. The reform era of the 1980s and 1990s put a premium on economic development in coastal regions, and telecommunications officials were quick to provide new telecommunications technologies to regions that could fuel the most rapid economic growth.

By the late 1990s, new technologies, such as wireless services and data networks, made equitable access more feasible. New uses of the Internet for educational, health, and other purposes made it more socially and economically desirable as a tool for narrowing gaps in income and standards of living. The Ministry of Education's CERNET project had been an effective though limited tool to further some of these goals. However, in the larger telecommunications arena, rural and poor areas continued into the following decade to be revenue losers, inspiring Ministry of Information Industry bureaucrats to maintain their resistance to subsidized expansion of the network to less-advantaged citizens.

It was unlikely the government's telecommunications officials would seek to bridge the digital divide for altruistic reasons. China's 'have nots' lacked the kind of political clout one might find in a democratic system. They were therefore left dependent on the will of bureaucrats who placed the development of their own economic sector above that of the interests of those who may have had a desire for technology, but who lacked the resources to acquire them.

As this chapter has indicated, the more basic technology of fixed-line telephone connections had started to penetrate to rural areas by the mid-2000s, and revenue losses had begun to narrow. Progress on the spread of Internet access to the countryside had been glacial though steady. It seemed reasonable to assume that, as urban and developed areas became more saturated with wireless communications and data network access, the inland areas would become the new sources of profit for officials guiding the telecommunications sector. Lower cost technologies and bureaucratic interests could then shift the technology's trajectory toward more equitable provision of communications access.

It was unlikely that government officials would soon relinquish control of China's telecommunications tools. But new technological and market forces acting on the sector betokened a slow but steady shift in bureaucratic emphasis toward those areas lacking advanced communications tools. Unless these changes occurred at a rapid enough pace, however, the economic, educational, and health divides, fostered in part by the digital divide, stood to lead to growing resentment and perhaps even unrest in the poor, inland parts of China.

8
Industrial Policy and Lessons of the Telecommunications Revolution

The first chapter of this study discussed the basic elements of industrial policy theory, and pointed out the potential pitfalls nations could face under state-guided economic development. Chalmers Johnson and his successor analysts found instances in which industrial policy could have positive effects, and this text has sought to add empirical evidence in a key sector of China's industrial development. From the case of China's telecommunications industry, we may add some corollaries to the core of state-guided intervention.

1. GOVERNMENTAL INTERVENTION WILL SUCCEED ONLY IF IT SERVES TO SATISFY A CLEAR LONG-TERM PUBLIC NEED.

As the previous chapters have outlined, the telecommunications sector brings a wide range of benefits to any nation's economic and social well-being. Communications tools at the basic level of the telephone allow even the poorest rural residents knowledge of markets and information to help improve their lot. Massive government spending in the 1980s and 1990s to build the fixed-line network, as discussed in Chapter 2, gave the majority of China's citizens at least limited access to modern communications systems.

At the other end of the spectrum, the spread of high-speed Internet access contributed to national and international commerce, as well as rapid exchange of economic, cultural, and political information. Investment in the basic tools of telecommunications equipment can also allow a developing nation like China to take a leading role in the international marketplace, and the success of companies like Huawei and ZTE testifies to the viability of the sectoral focus.

Lessons of the Revolution

This is not to say that all of China's industrial policies have been successes. Chapter 1 noted several other areas of industrial development in which the long-term public need for the industry was not so clear. Among these, the steel industry and even automobile production stand out as questionable, at least in the way the government pursued their development. From almost the beginning of China's reform era, however, as Chapter 2 indicated, telecommunications was perceived as a key area with a broad array of benefits.

2. GOVERNMENTAL INTERVENTION SHOULD INCLUDE THE ENCOURAGEMENT OF VIGOROUS COMPETITION IN ORDER TO SUSTAIN VITALITY AND UNDERMINE BUREAUCRATIC SLOTH. HOWEVER, THE INTERVENTION SHOULD ALSO INSIST ON MAINTAINING SUFFICIENT GOVERNMENTAL CONTROL TO ASSURE THAT LONG-TERM GOALS ARE STEADFASTLY PURSUED—NOT FORGOTTEN OR UNDERMINED IN THE DRIVE FOR FINANCIAL SUCCESS.

In the many parts of China's telecommunications considered in this study, those where competition prevailed advanced most quickly. Chapter 3 illustrated this in the case of the rise of the mobile phone network; Chapter 4 showed the growth of Internet content and electronic commerce companies; and Chapter 5 chronicled the competing equipment manufacturers. In the case of the mobile phone system and equipment makers, public companies played a key role in the competition; for Internet content, the government left development to the private sector. But even where state ownership dominated, the government allowed and encouraged competition to raise quality and lower prices, and spread the technologies so a broad sector of the population could see benefits.

For a developing country such as China, state intervention was in most cases necessary to maintain momentum. Had the industries been left to the domestic private sector, a lack of capital may have retarded growth (particularly in the case of building a nationwide mobile phone network). An early and unguided opening to foreign investment may have led to incoherent development under the guidance of disparate entrepreneurs focused on a quick return on investment. As Chapter 2 indicated, developing nations such as India and countries in Africa suffered as the state failed to play a leading role to mobilize capital to build nationwide networks, and old-line government ministries served as barriers to progress.

3. PURELY NATIONALISTIC TENDENCIES SHOULD GIVE WAY TO THE NEED TO BRING IN MORE ADVANCED FOREIGN PARTNERS FROM WHOM TO LEARN TO ACHIEVE WORLD CLASS COMPETITIVENESS AND AVOID MISTAKES ONE WOULD MAKE ON ONE'S OWN.

In many arenas of China's development over the past twenty-five years, foreign investment has played a key role. This makes the PRC a sharp contrast with the cases of Japan and South Korea, and, in the developing world, India, which have pursued paths designed in many ways to avoid foreign equity participation in domestic enterprises.

Despite several decades of isolation and distrust of Western economic powers, the PRC was willing to open to foreign participation in key industrial sectors. It did not, however, allow the foreign investors a dominant position. As Chapter 3 (for mobile phones), Chapter 4 (on Internet content), and Chapter 5 illustrated, the government kept limits on foreign ownership, ones that allowed the country to enjoy the higher quality of joint-venture production, while at the same time absorb technology and learn foreign methods of management and production. The country was further able to 'leapfrog' the necessity of developing its own brands of network construction and management, and move quickly to an international standard of production and service.

The caveat to this point is that a country should maintain vigilance to give away only so much of the rewards as will entice willing partners to join, without jeopardizing the nation's future competitiveness on an international stage. The ability of companies such as Huawei, ZTE, and probably other future companies to succeed beyond China's borders indicate the state achieved this delicate balance for the telecommunications equipment and perhaps telecommunications services sectors.

4. ALL ALONG, CLOSE TIES BETWEEN NATIONAL AND LOCAL BUREAUCRACIES ARE NECESSARY TO MAINTAIN AN OVERARCHING POLICY AND A COORDINATED APPROACH.

This point is particularly valid for large, developing nations where municipal or regional governments have a significant degree of economic autonomy. For typical nations that have employed industrial policy to guide development, such as Japan and South Korea, the national government held a tighter grip on financial and regulatory tools. Major cities or prefectures in Japan play a relatively small role in furthering the goals of economic growth.

Chapter 6 illustrated the major role the Shanghai city government played in orchestrating nearly all aspects of telecommunications advance in

Lessons of the Revolution

China. The chapter also chronicled that cities and provinces throughout the nation have similar powers to shape and encourage economic projects within their own limited boundaries. For telecommunications, Shanghai harmonized its plans with national goals to achieve rapid and useful results.

A counter-example to the Shanghai case is found in India. Peter Evans used the case of the Indian state of Kerala to show how the regional government, led by a Communist Party, initially had great success in instilling some of the nation's highest levels of literacy, life-expectancy, and general social well-being.[1] A program to redistribute land from an old landlord class to peasant proprietors inspired a revolution in agricultural production. Local government agencies were key in the 1960s and 1970s for providing health-centers, subsidized food, and education facilities. Indian civil servants were vital to running these services.

Despite its success in building its own local agricultural and social-welfare infrastructure, the regional Kerala government was less capable of inspiring the creation of a class of industrial entrepreneurs. By the 1980s, the local economy had begun to stagnate, just as the national government was embarking on a series of economic changes to revitalize the economy.

In contrast to Kerala, the regional government for the Indian city of Bangalore played an important role in developing the city as a major target for foreign technology investment.[2] Before national liberalization policies began in 1991, the state government had focused public investment in companies engaged in telecommunications, aeronautics, machine tools, and other fields. In the 1990s and early 2000s, foreign companies set up manufacturing facilities to take advantage of the high-technology core.

Despite Bangalore's seeming success, both the regional and national government failed the new city in many regards. Key infrastructure facilities such as electricity, water, roads, air ports, and a host of other supporting features were of poor quality even several years after the boom began. Therefore, unless government at all levels presents a coordinated effort to coherently nurture and support a targeted sector, it is likely the industry will face developmental difficulties.

5. A KEY PROBLEM IN FURTHERING NATIONAL INITIATIVES IS KEEPING CORRUPT PRACTICES TO A MINIMUM. THIS IS A DYNAMIC PROBLEM BECAUSE EACH NEW MEASURE TO KEEP CORRUPTION IN CHECK CAN BE MET BY STEPS TO CIRCUMVENT IT.

In the developmental cases chronicled in this text, instances of blatant corruption seemed to stay at a minimum. Jiang Mianheng, the son of

Chinese leader Jiang Zemin, accumulated several titles in the telecommunications sector. He seemed to be taking advantage of political connections to be part of the rapidly growing industry. As of the mid-2000s, however, he had been accused of no particular crimes, even as his father had relinquished most of his political positions.

The case of Huawei also included intimations of illicit dealings, as the company created gray-area joint ventures with local phone companies, ones which seemed to be channeling profits to those buying equipment from Huawei. The company moved within a few years to dismantle the ventures, perhaps following government pressure. Moreover, Huawei's ability to export indicates that the company's equipment quality was as good as or at least near that of competitors, so that its domestic sales may still have contributed to a healthy communications network.

China does have a reputation as a nation troubled by corruption. In 2004, the monitoring group Transparency International ranked the country seventy-first out of 146 countries (with 146 being the most corrupt); the same year, India ranked ninetieth.[3] China also had the custom of prosecuting only those receiving bribes, not offering them, and people who received kickbacks from sales transactions were also not liable. Still, in early 2006, the government launched an anti-corruption drive, and the Central Commission for Discipline Inspection moved to publicize its efforts to root out illicit behavior.[4] The result of efforts to combat the problem was an improved ranking for China by Transparency International from 1995 to 2004; India's raw score remained static over the same decade.

6. A NATIONAL POLICY SHOULD AIM FIRST TO SOLVE THE MOST READILY ADDRESSABLE PROBLEMS, AND LEAVE THE MORE DIFFICULT ONES FOR LATER.

In the case of telecommunications, this axiom led China into a digital divide. Citizens in wealthy parts of the nation benefited, while those in poorer, inland regions lagged behind. However, as Chapters 3 and 4 noted, those citizens who were early recipients of either telephones or Internet service paid a premium in connection and service fees. These payments helped lower costs in future years for succeeding waves of subscribers. As Chapter 7 indicated, newer technologies could lead to even lower costs to extend services to those currently not served by public communications networks.

In a larger sense, while the strategy seems obvious, it is difficult to implement in nations where citizens have a greater democratic voice in state allocation of funds. For China, it is possible inland rural citizens would have vocally opposed building a network that in its early years targeted mainly coastal citizens. In this case, the free hand given the national bureaucracy enabled it to construct a network that has had profoundly positive effects on a broad array of economic advances, ones that extend to the nation as a whole.

This part of China's industrial policy path is therefore one of the most difficult to replicate outside of nations where the state dominates political life. Most other developing nation governments with contending political parties vying for power, such as those in India and other east and southeast Asian countries, would likely face challenges if they focused attention on one region of the country before addressing the needs of the majority of citizen voters.

7. AS A DEVELOPING NATION ASCENDS AN INTERNATIONAL STAGE AND TAKES ON GLOBAL ECONOMIC RESPONSIBILITIES, THE COUNTRY FACES NEW PROBLEMS REQUIRING NEW STRATEGIES TO ASSURE THE CONTINUING PROSPERITY OF ITS CITIZENS IN THE FACE OF NEW COMPETITIVE FORCES.

For China, entry to the WTO in 2001 introduced both challenges and opportunities for its telecommunications industrial policy. As earlier chapters noted, China was able to limit foreign ownership in several of the fundamental telecommunications services areas. For example, even several years after the country's entry to the trade body, there were no major foreign-invested projects to operate fixed-line or mobile phone services.

In some areas, however, China did open more quickly. Chapter 4 pointed out the rapidly growing role of foreign companies in the Internet business, as non-Chinese companies took equity stakes in domestic companies, or expanded their businesses via an online presence. Chapter 5 showed that major joint-venture equipment makers could even acquire majority ownership, and felt greater protection under WTO intellectual property rights rules. Such policies also conform with the assertions of Pack and Saggi (presented in Chapter 1): namely, that foreign firms can provide the beneficial function of facilitating cost reduction in specific sectors.

In the mid-2000s, China was also in a position to leverage trade-related issues to protect key areas such as telecommunications. For example, it could pledge to limit textile exports to countries feeling swamped by Chinese goods in exchange for reduced foreign pressure to open access to its mobile phones operations, or to choose a new generation mobile phone standard more beneficial to foreign technology corporations.

As technologies within the country caught up to foreign-quality levels, domestic companies stood to take advantage of more liberal global trade rules. For Huawei, ZTE, and even operating companies like China Mobile, the opportunity to export their products and services to developing and even developed nations was facilitated by a global realization that free trade had tangible benefits.

Other developing nations could follow China's example by choosing vital sectors for protection, limiting foreign investment in ways that allow technology absorption and a reasonable financial gain for foreign companies, while not stifling nascent industries that could eventually become themselves globally competitive. China, of course, was blessed with a large domestic market, one that could offer a profit prize while extracting foreign know-how. Nations with smaller markets, ones that are targets for neither foreign investment manufacture nor even perhaps domestic entrepreneurs, could face strong pressure to open to foreign-made goods to satisfy internal needs.

In a broader sense, then, the role that governments need to play to succeed as WTO liberalization continues was not clear in the first decade of the twenty-first century. The continued integration of more developing nations into the global trade regime stood to bring fresh empirical evidence as more nations opened their economic borders to flows of goods and services in both directions.

China's telecommunications revolution proved by nearly all measures a spectacular success. The role of the state was vital and positive, and gave hundreds of millions of the country's citizens broad access to cutting-edge communications tools.

The present study began with a summary of industrial policy theories of government intervention and sought to determine the extent to which these ideas were applicable to China's telecommunications development. The points listed above should be viewed as corollaries that specify conditions necessary for the ideas of Chalmers Johnson and other kinship theorists to apply. They provide a kind of road map for nations seeking to emulate China's success, not just in the telecommunications field, but also in other aspects of economic development. If the area chosen for state intervention is well considered, competition is encouraged, and global economic forces are skillfully harnessed, other nations can emulate such economic 'revolutions' in a broad array of sectors.

Notes

Chapter 1

1. Chalmers Johnson, *MITI and the Japanese Miracle*. Stanford, Calif.: Stanford University Press, 1982: 18.
2. Ibid. 19. A World Bank study has a similar definition of 'industrial policy,' calling it 'government efforts to alter industrial structure to promote productivity-based growth.' (World Bank, *The East Asian Miracle: Economic Growth and Public Policy*. New York: Oxford University Press, 1993: 304.
3. For more detail on these and other elements of Johnson's industrial policy model, see Johnson, *MITI and the Japanese Miracle*, 318–19, as well as his entire concluding chapter (pp. 305–24).
4. Ibid. 276–8.
5. Ibid. 314.
6. Ibid. 309.
7. Ibid. 310.
8. Ibid.
9. Stephan Haggard and Tun-jen Cheng, 'State and Foreign Capital in the East Asian NICs'. In Frederic C. Deyo, ed. *The Political Economy of the New Asian Industrialism*. Ithaca, NY: Cornell University Press, 1987: 84–129.
10. Ibid. 117.
11. Ezra Vogel, *The Four Little Dragons: The Spread of Industrialization in East Asia*. Cambridge, Mass.: Harvard University Press, 1991.
12. Ibid. 95.
13. World Bank, *The East Asian Miracle*.
14. Ibid. 349–52.
15. Ibid. 314–15. For a definition of TFP, see ibid. 54–5. Essentially, it refers to output growth minus 'the portion of growth due to capital accumulation, to human capital accumulation, and labor force growth'.
16. Ibid. 326.
17. Sanjaya Lall, *Learning from the Asian Tigers: Studies in Technology and Industrial Policy*. Houndmills: Macmillan, 1996: 121.
18. Ibid. 120.
19. Ibid. 103–5.
20. Ibid. 209.

21. Peter Evans, 'Class, State, and Dependence in East Asia: Lessons for Latin Americanists.' In Deyo, *The Political Economy of the New Asian Industrialism*: 222, 206.
22. Ibid. 218.
23. Ibid. 215.
24. Ibid. 223.
25. Marcela Miozzo, 'Transnational Corporations, Industrial Policy and the "War of Incentives": The Case of the Argentine Automobile Industry', *Development and Change* 31 (2000): 651–80.
26. Lall, *Learning from the Asian Tigers*, ch. 5, pp. 124–47; ch. 7, pp. 166–96.
27. Ibid. 76–7.
28. Paul Krugman, 'The Myth of Asia's Miracle'. *Foreign Affairs* 73 (1994): 62–78; Seiichi Masuyama, et al., eds., *Industrial Policies in East Asia*. Tokyo: Nomura Research Institute, 1997.
29. Masuyama, 'The Evolving Nature of Industrial Policy in East Asia: Liberalization, Upgrading, and Integration'. in id., *Industrial Policies in East Asia*: 4.
30. *Businessweek*, 22 Dec. 1997, p. 24.
31. Ibid.
32. François Godemont, *The Downsizing of Asia*. New York: Routledge, 1999: 201.
33. Ibid. 23–4.
34. Ibid. (see ch. 2).
35. Ibid. (see ch. 8).
36. *The Economist*, 17 May 2003, p. 90.
37. Ibid. var. year reports on GDP growth.
38. Peter Evans, *Embedded Autonomy: States and Industrial Transformation*. Princeton, NJ: Princeton University Press, 1995: 12.
39. Ibid.
40. Howard Pack and Kamal Saggi, 'Is There a Case for Industrial Policy? A Critical Survey', *The World Bank Research Observer* 21 (2006): 268.
41. Ibid. 292–3.
42. Andrew P. Cortell, *Mediating Globalization*. Albany, NY: State University of New York Press, 2006: 190.
43. Ibid. 194.
44. Dorothy Solinger, *From Lathes to Looms: China's Industrial Policy in Comparative Perspective, 1979–1982*. Stanford, Calif.: Stanford University Press, 1991: 30–6.
45. Ibid. 235.
46. Ibid. 221.
47. Ibid. 123–93.
48. Kenneth Lieberthal and Michel Oksenberg, *Policy Making in China: Leaders, Structures, and Processes*. Princeton, NJ: Princeton University Press, 1988.
49. Edward S. Steinfeld, *Forging Reform in China: The Fate of State-owned Industry*. Cambridge: Cambridge University Press, 1998.

50. Thomas G. Moore, *China in the World Market: Chinese Industry and International Sources of Reform in the post-Mao Era*. Cambridge: Cambridge University Press, 2002.
51. Adam Segal, *Digital Dragon: High Technology Enterprises in China*. Ithaca, NY: Cornell University Press, 2003.
52. Daniel C. Lynch, *After the Propaganda State: Media, Politics, and 'Thought Work' in Reformed China*. Stanford, Calif.: Stanford University Press, 1999. On telecommunications, see in particular pp. 165–75 and pp. 208–15.
53. Eric Harwit, *China's Automobile Industry: Policies, Problems, and Prospects*. Armonk, NY: M. E. Sharpe, 1995.
54. Steinfeld, *Forging Reform in China*, 5.
55. Ibid. 254.
56. Ibid. 255.
57. Moore, *China in the World Market*, 302.
58. Segal, *Digital Dragon*, 7.
59. Milton Mueller and Zixiang Tan, *China in the Information Age*. Westport, Conn.: Praeger Publishers, 1997.
60. Ding Lu and Chee Kong Wong, *China's Telecommunications Market: Entering a Competitive Age*. Cheltenham: Edward Elgar, 2003.
61. Xu Yan and Douglas Pitt, *Chinese Telecommunications Policy*. Boston, Mass.: Artech House, 2002.
62. Zixiang Tan, 'China's Information Superhighway' What Is It and Who Controls It?' *Telecommunications Policy* 19 (1995): 721–31; Wei Wu, 'Great Leap or Long March': Some Policy Issues of the Development of the Internet in China'. ibid. 20 (1996): 699–711; Zixiang Tan, 'Regulating China's Internet: Convergence Toward a Coherent Regulatory Regime'. ibid. 23 (1999): 261–76.
63. Bruce T. McIntyre, 'China's Use of the Internet: A Revolution on Hold.' In Paul S. N. Lee, ed. *Telecommunications and Development in China*. Cresskill, NJ: Hamption Press, 1997: 149–69.
64. Geoffry Taubman, 'A Not-So World Wide Web: The Internet, China, and the Challenges to Nondemocratic Rule'. *Political Communication* 15 (1998): 255–72.
65. Daniel C. Lynch, 'Dilemmas of "Thought Work" in *Fin-de-Siècle* China'. *China Quarterly* 157 (1999): 173–201.
66. Michael Chase and James Mulvenon, *You've Got Dissent! Chinese Dissidents Use of the Internet and Beijing's Counter-strategies*. Santa Monica, Calif.: RAND, 2002.
67. Shanhi Kalathil and Taylor Boas, 'Wired for Modernization in China'. In eid., *Open Networks, Closed Regimes: The Impact of the Internet on Authoritarian Rule*. Carnegie Endowment for International Peace, Jan. 2003: 13–42.
68. Xiangdong Wang, *Xinxihua: Zhongguo 21 shiji de xuanze* (Informatization: China's Choices in the 21st century). Beijing: Shehuikexue wenzhai chubanshe, 1998.

69. Richard Katz, *Japan, the System that Soured: The Rise and Fall of the Japanese Economic Miracle*. Armonk, NY: M. E. Sharpe, 1998.
70. Ibid. 32–3.
71. Adam Segal and Eric Thun, 'Thinking Globally, Acting Locally: Local Governments, Industrial Sectors, and Development in China'. *Politics and Society* 29 (2001): 557–88.

Chapter 2

1. George P. Oslin, *The Story of Telecommunications*, Macon, Ga.: Mercer University Press, 1992: 73, 83, 198.
2. Ibid. 241–2.
3. Ibid. 221, 222.
4. Ibid. 228, 230; Richard E. Wiley, 'Competition and Deregulation in Telecommunications: The American Experience', in Leonard Lewin, ed. *Telecommunications in the U.S.: Trends and Policies*, Dedham, Md.: Artech House, 1981: 39.
5. Wiley, 'Competition and Deregulation in Telecommunications', 39–40.
6. Ibid. 40.
7. John Brooks, *Telephone: The First Hundred Years*, New York: Harper & Row, 1975: 133–6.
8. Wiley, 'Competition and Deregulation in Telecommunications', 39, 41.
9. Clyde T. Ellis, *A Giant Step*, New York: Random House, 1966: 108.
10. Ibid.
11. Ibid. 109.
12. Wiley, 'Competition and Deregulation in Telecommunications', 41.
13. Eli Noam, *Telecommunications in Europe*, New York: Oxford University Press, 1992: 18, 22.
14. Ibid. 103–6.
15. Ibid. 134–7.
16. Ibid. 137.
17. Ibid. 141–2.
18. Parts of the following discussions on Latin America, Africa, and India are based on Eric Harwit, 'China's Telecommunications Industry: Development Patterns in Policies', *Pacific Affairs* 71 (1998): 177–8.
19. This discussion is based on Ben A. Petrazzini, *The Political Economy of Telecommunications Reform in Developing Countries*, Westport, Conn.: Praeger, 1995: 107–11.
20. Latin America News Digest, 3 Feb. 2005 report.
21. Petrazzini, *The Political Economy of Telecommunications Reform*, 49–102.
22. *New York Times*, 23 Apr. 1990, p. D8.
23. United Press International, 5 Oct. 1990 report.
24. IPS Interpress Service, 11 Aug. 2003 report.

25. Information on Africa to the 1990s is derived from Raymond U. Akwule, 'Telecommunciations in Africa: Policy and Management Trends', in Meheroo Jusawalla, ed. *Global Telecommunications Policies*, Westport, Conn.: Greenwood Press, 1993: 159–70.
26. AllAfrica, Inc., Africa News, 14 Jan. 2004 report.
27. Jarice Hanson and Uma Narula, *New Communication Technologies in Developing Countries*, Hillsdale, NJ: Lawrence Erlbaum Associates, 1990: 50.
28. C. S. Swaminathan, 'Role of Telecommunications in National Economy', in *Telecommunications for National Development*, Calcutta: India Chamber of Commerce, 1983: 83.
29. Hanson and Narula, *New Communication Technologies*, 51–2.
30. *Far Eastern Economic Review*, 7 Apr. 1994, p. 44.
31. India's Ministry of Commerce and Industry website gave more detail on how foreign corporations could invest in the nation. See http://www.dipp.nic.in/.
32. Financial Times Information, *Indian Express*, 19 Nov. 2003.
33. Many history texts discuss this period of Japan's development; see e.g. Edwin O. Reischauer and Albert Craig, *Japan: Tradition and Transformation*, Boston, Mass.: Houghton Mifflin, 1989, in particular ch. 5 (pp. 145–89).
34. E. Sydney Crawcour, 'Industrialization and Technological Change, 1885–1920', in Kozo Yamamura, ed. *The Economic Emergence of Modern Japan*, Cambridge: Cambridge University Press, 1997: 64; Reischauer and Craig, *Japan*, 148.
35. Éva Ehrlich, *Japan: A Case of Catching Up*. Budapest: Akadémiai Kiadó, 1984: 133.
36. Marie Anchordoguy, 'Nippon Telegraph and Telephone Company (NTT) and the Building of a Telecommunications Industry in Japan', *Business History Review* 75 (2001): 510.
37. Clare Doran and Christina M. Stansell, 'Nippon Telegraph and Telephone Corporation', in Tina Grant, ed. *International Directory of Company Histories* 51. Detroit, Mich.: St James Press, 2003: 271.
38. Anchordoguy, 'Nippon Telegraph and Telephone Company', 511–12.
39. Ehrlich, *Japan*, 178 n. 112.
40. Harumasa Sato and Rodney Stevenson, 'Telecommunications in Japan', *Columbia Journal of World Business* 24 (1989): 31. Marie Anchordoguy points out that many of the MPT regulators were in fact specialists from NTT, so that NTT in effect regulated itself. See Anchordoguy, 'Nippon Telegraph and Telephone Company', 517.
41. Ehrlich, *Japan*, 180 n. 118.
42. James F. Larson, *The Telecommunications Revolution In Korea*, Hong Kong: Oxford University Press, 1995: 33.
43. Ibid. 117.
44. Ibid. 36.
45. Andrew Harrington, 'Companies and Capital in Asia–Pacific Telecommunications', in John Ure, ed. *Telecommunications in Asia*, Hong Kong: Hong Kong University Press, 1997: 85.

46. The following summary argument is derived from Harwit, 'China's Telecommunications Industry', 178–9.
47. See Erik Baark, *Lightning Wires*, Westport, Conn.: Greenwood Press, 1997. Other early history of the industry are found in *Zhongguo jindai youdian shi* (Modern History of China's Posts and Telecommunications), Beijing: Renmin youdian chubanshe, 1984, esp. chs. 5–8; *Dangdai zhongguo de youdian shiye* (Contemporary China's Posts and Telecommunications Facilities), Beijing: Dangdai zhongguo chubanshe, 1993: 9–18; and in *Zhongguo jiaotong nianjian* (China Communications Yearbook), Beijing: Zhongguo jiaotong nianjian she, 1986: 28.
48. Baark, *Lightning Wires*, 73.
49. Ibid. 74.
50. Ibid. 75, 77.
51. Ibid. 77–86.
52. Ibid. 110–12.
53. Ibid. 130–2.
54. Ibid. 163–6, 177.
55. Albert Feuerwerker, *China's Early Industrialization*, Cambridge, Mass.: Harvard University Press, 1958: 192–207. Unless otherwise noted, the following discussion of the growth of telegraph lines in the late nineteenth century is based on this account.
56. Ibid. 205–6.
57. Ibid. 207.
58. Ibid.
59. Ibid. 80.
60. Early telephone use is discussed in Shu-hwai Wang, 'China's Modernization in Communications', in Chi-ming Hou and Tzong-shian Yu, ed. *Modern Chinese Economic History*, Taipei: Institute of Economics, Academia Sinica, 1977: 335–51. The following discussion is derived from this chapter.
61. The following details on China's communications network in the 1920s and 1930s is derived from Leang-li T'ang, ed., *Reconstruction in China*, Shanghai: China United Press, 1935: 278–87.
62. Under the Nationalists, the administration of postal affairs was put into a Directorate-General of Posts, under the Ministry of Communications. Ibid. 269.
63. Ibid. 278.
64. Ibid. 279.
65. Ibid. 280.
66. *Zhongguo jindai youdian shi* (Modern History of China's Posts and Telecommunications): 186.
67. T'ang, *Reconstruction in China*, 279–81.
68. *Dangdai zhongguo de youdian shiye* (Contemporary China's Posts and Telecommunications Facilities): 13–14.
69. *Statistical Yearbook of China, 1981* (English Edn.), Hong Kong: Economic Information & Agency, 1982: 295.

70. *Zhongguo jindai youdian shi* (Modern History of China's Posts and Telecommunications): 225.
71. *Zhongguo jiaotong nianjian* (China Communications Yearbook), 1986: 28.
72. The ideas in this paragraph are based on a discussion in Harwit, 'China's Telecommunications Industry', 180.
73. For detail on this type of communication, see Frederick T. C. Yu, 'Communications and Politics in Communist China', in Lucian Pye, ed. *Communications and Political Development*, Princeton, NJ: Princeton University Press, 1963: 259–97.
74. Reference to Mao's comments in 1956 were published in 1977 in Xinhua General Overseas News Service, 21 Sept. 1977; this is a citation of an article originally published in *People's Daily*, 15 Sept. 1997, p. 1.
75. Duanquan Zhang, 'Development and Modernization of Telecommunications in New China', *China Telecommunications Construction* 1 (1989): 4.
76. Yuan-li Wu, *An Economic Survey of Communist China*, New York: Bookman Associates, 1956: 323.
77. Information on the early planning system of telecommunications manufacturing is derived from *Dangdai zhongguo de youdian shiye* (Contemporary China's Posts and Telecommunications Facilities): 514–19.
78. Doak A. Barnett, *Cadres, Bureaucracy, and Political Power in Communist China*, New York: Columbia University Press, 1976: 80–1.
79. Ibid. 305.
80. Ibid. 72–3.
81. For further discussion of the tiao-kuai dichotomy, see ibid. 73; Audrey Donnithorne, *China's Economic System*, New York: Praeger, 1967: 152; and, in more recent years, Kenneth Lieberthal, *Governing China: From Revolution Through Reform*, New York: Norton, 1995: 169–70.
82. Wolfgang Bartke, *Who was Who in the People's Republic of China*, Munich: K. G. Saur Verlag, 1997: 695; Elizabeth Perry, *Shanghai on Strike*, Stanford, Calif.: Stanford University Press, 1993: 96–101.
83. Donnithorne, *China's Economic System*, 153.
84. Jan Prybyla, *The Chinese Economy*, Columbia, SC: University of South Carolina Press, 1978: 197.
85. Jack Craig, 'China: Domestic and International Telecommunications, 1949–1974', in Joint Economic Committee, Congress of the United States, *China: A Reassessment of the Economy*, Washington, DC: US Government Printing Office, 1975: 293.
86. *Zhongguo youdian baike quanshu* (Encyclopedia of China's Posts and Telecommunications), Beijing: Renmin youdian chubanshe, 1995: 728; Craig, 'China: Domestic and International Telecommunications,' 294; Bartke, *Who was Who*, 670.
87. Duanquan Zhang, 'Development and Modernization of Telecommunications in New China', 4.
88. The ideas in this para. are based on a discussion in Harwit, 'China's Telecommunications Industry', 181, 183.

89. Barry Richman, *Industrial Society in Communist China*, New York: Random House, 1969: 584–5.
90. Cited in Xinhua General Overseas News Service, 21 Sept. 1977; original article in *People's Daily*, 15 Sept. 1997, p. 1.
91. *People's Daily*, 15 Sept. 1997, p. 1, author's translation.
92. The ideas in this paragraph are based on a discussion in Harwit, 'China's Telecommunications Industry,' 183.
93. Bartke, *Who was Who*, 491–2, 499.
94. From Barry Naughton, *Growing Out of the Plan: Chinese Economic Reform, 1978–1993*, New York: Cambridge University Press, 1995: 108. Naughton's text gave a thorough discussion of the Chinese government's movements away from political control and toward rational market mechanisms.
95. Duanquan Zhang, 'Development and Modernization of Telecommunications in New China', 11.
96. Information on Yang's background is derived from *Who's Who in China Current Leaders*, 783–4.
97. Dacai Gong, ed., 'Youdian tongxinye' (Post and Telecommunications Industry), in Hong Ma and Shangqing Sun, eds. *Xiandai zhongguo jingji da shi dian* (Compendium on Modern China's Economy), Beijing: Zhongguo caizheng jingji chubanshe, 1993: 1844–6.
98. John Ure, 'Telecommunications in China and the Four Dragons', in id., *Telecommunications in Asia*, 16.
99. Dacai Gong, 'Youdian tongxinye' (Post and Telecommunications Industry), 1844.
100. Ure, 'Telecommunications in China', 16; *Zhongguo tongji nianjian* (China Statistical Yearbook), Beijing: Zhongguo tongji chubanshe, 2004: 662.
101. Hong Ma and Shangqing Sun, 'Youdian tongxinye zhongda fangzhen zhengce' (Major Principles and Policies of the Post and Telecommunications Industry), *Xiandai zhongguo jingji da shi dian*, 1853. See also Ure, 'Telecommunications in China', 16.
102. Ure, 'Telecommunications in China', 16.
103. Ibid. 17.
104. *Zhongguo tongji nianjian* (China Statistical Yearbook), 2004: 659–60 and 662. Of course, many mobile phone users also had fixed-line phones in their homes and offices, so the actual penetration of telephones into Chinese society is lower than the teledensity figure would imply.
105. Andrew Harrington, 'Companies and Capital in Asia–Pacific Telecommunications', 85.
106. Based on total number of subscribers listed in Xinhua Economic News Service, 6 July 2007 report.
107. Interview with Chinese telecommunications scholar in Beijing, 1996.
108. *South China Morning Post*, 28 Aug. 1997, China Business Review section, p. 2. The exchange rate used in this text for events from 1995 to 2005 is about US dollar 1= Chinese yuan 8.3. In previous and succeeding years, the rate varied.

Notes to pp. 40–48

109. Information on Wu's background is derived from *Who's Who in China Current Leaders*, 697–8.
110. *The Economist*, 9 Dec. 2000, p. 76.
111. Ibid.
112. *Economist Intelligence Unit*, Business China, 15 Mar. 1999, p. 3.

Chapter 3

1. On changes in the years following the AT&T breakup, see *The Economist*, 5 Oct. 1991, p. 12.
2. Ibid. 30 May 1992, p. 19.
3. *Guardian* (London), 12 Apr. 2004, p. 22. For information on the break-up of British Telecom, see the company's history on the BT Group website, http://www.btplc.com/Thegroup/BTsHistory/History.htm.
4. From Hoover's, Inc. website. See http://www.hoovers.com/france-telecom/-ID__43307-/freeuk-co-factsheet.xhtml.
5. *The Economist*, 21 Aug. 2004.
6. Marie Anchordoguy, 'Nippon Telegraph and Telephone Company (NTT) and the Building of a Telecommunications Industry in Japan', *Business History Review* 75 (2001): 531.
7. Detail on development of telecommunications in Africa can be found in Raymond U. Akwule, 'Telecommunications in Africa: Policy and Management Trends', in Meheroo Jusawalla, ed. *Global Telecommunications Policies*, Westport, Conn.: Greenwood Press, 1993: 159–70.
8. The following discussion is based on Eli Noam, *Telecommunications in Europe*, New York: Oxford University Press, 1992: 55–60.
9. Ibid. 430.
10. Interview in 1996 in Beijing with Chinese telecommunications university professor who was part of the research team. Unless otherwise indicated, the following paragraphs on Unicom's early years are based on his account.
11. An official at the MPT, however, indicated World Bank influence was relatively small in the move to create Unicom. From 1996 interview in Beijing.
12. Daniel Lynch made this point in his book *After the Propaganda State: Media, Politics, and 'Thought Work' in Reformed China*, Stanford, Calif.: Stanford University Press, 1999: 166. Lynch also speculated that Hu may have been trying to 'get even' with the MPT for the telecommunications ministry's refusal to source equipment from MEI companies, and for discouraging multinational telecommunications equipment companies from cooperating with the MEI.
13. From 1996 interview, Beijing.
14. Lynch, *After the Propaganda State*, 167–8; *South China Morning Post*, 18 Feb. 1993, Business Post, p. 1.
15. The full name in Chinese is *Zhongguo Lianhe Tongxin Youxian Gongsi*, and in English China United Telecommunications Corporation. As noted later on, the

2000 public listing of 'China Unicom' delineated the parent company from the listed one, with the parent keeping the full English-language name.
16. Eric Harwit and Jack Su, 'A Telecom Newcomer Challenges the MPT Monopoly', *China Business Review* 23 (1996): 22; *USITO China IT Issue Report* 1 (1997): 4.
17. The Chinese military, which also had its own wireless communication system, was not one of Unicom's shareholders.
18. Harwit and Su, 'A Telecom Newcomer Challenges the MPT Monopoly', 22.
19. Unicom company brochure, 1995, p. 25.
20. *South China Morning Post*, 20 July 1994, Business Post, p. 1.
21. Unicom brochure, 27.
22. Lynch, *After the Propaganda State*, 169.
23. China Online, 3 Dec. 2001 report.
24. Xu Yan and Douglas Pitt, *Chinese Telecommunications Policy*, Boston, Mass.: Artech House, 2002: 19.
25. 'Guanyu Jiejue Liantong Gongsi Fazhan Wenti de Diaocha Baogao' (A survey findings report on solving the Unicom company's development problems). Internal memo from the State Economic and Trade, Planning, and State System Restructuring Commissions to then vice premier Zou Jiahua and then vice premier Wu Bangguo, 14 Mar. 1997, pp. 8–9.
26. Kyodo News Service, 29 July 1995 report.
27. BDAssociates, Ltd., *China Telecom Update*, 3.
28. *USITO China IT Issue Report*, 1 (1997): 8.
29. 'Guanyu Jiejue Liantong Gongsi Fazhan Wenti de Diaocha Baogao' (A survey findings report on solving the Unicom company's development problems), 3. See also Harwit and Su, 'A Telecom Newcomer Challenges the MPT Monopoly', 22–3.
30. From July 1997 interview with Shanghai Unicom official.
31. *South China Morning Post*, 2 June 1998, Business Post, p. 4.
32. *Asia Pacific Telecoms Analyst* report, 2 June 1997.
33. 'Guanyu Jiejue Liantong Gongsi Fazhan Wenti de Diaocha Baogao' (A survey findings report on solving the Unicom company's development problems), 11.
34. *USITO China IT Issue Report*, 8.
35. Interview with Unicom official in Beijing, May 1996.
36. Economist Intelligence Unit, *Business China*, 24 July 1995, p. 2.
37. 'Guanyu Jiejue Liantong Gongsi Fazhan Wenti de Diaocha Baogao' (A survey findings report on solving the Unicom company's development problems), 6.
38. From June 1996 interviews with Ameritech and Bell Canada.
39. *USITO China IT Issue Report*, 7.
40. Interview with Unicom official in Beijing, May 1995.
41. From interview series in China, 1996 and 1997.
42. *South China Morning Post*, 11 Jan. 2000, Business Post, p. 1.
43. *Asia Pulse*, 1 Sept. 1997 report.
44. Unless otherwise noted, the following information on Ameritech comes from a June 1996 interview with an Ameritech representative in Beijing.

45. *Financial Times*, 19 Mar. 1997, Telecommunications Review, p. 3.
46. *South China Morning Post*, 11 Aug. 1997, Business Post, p. 1. A joint venture of Hong Kong's CCT Telecom Holdings, the American company McCaw Telecommunications System, and the Shanghai city government later took an 80 percent stake in a joint venture in Shanxi, apparently replacing Ameritech. See ibid. 11 Jan. 2000, Business Post, p. 1.
47. Ibid. 7 May 1997, Business Post, p. 4.
48. From summer 1996 interview in Beijing.
49. Ibid.
50. 'Guanyu Jiejue Liantong Gongsi Fazhan Wenti de Diaocha Baogao' (A survey findings report on solving the Unicom company's development problems), 1–21.
51. Ibid. 17–18. Author's translation, emphasis added.
52. Ibid. 11–12.
53. Ibid. 12.
54. Ibid. 12, 15.
55. Ibid. 16–20.
56. Ibid. 7.
57. *South China Morning Post*, 7 May 1997, Business Post, p. 4.
58. From interview with foreign consultant based in Beijing, 1996.
59. Ibid. and *Telecommunication Development Asia-Pacific*, Dec. 1995, p. 52.
60. *South China Morning Post*, 28 Aug. 1997, China Business Review, p. 10.
61. *USITO China IT Issue Report*, 5.
62. Ibid. 3.
63. *Straits Times* (Singapore), 24 July 1996, Money 4, p. 37.
64. *South China Morning Post*, 7 May 1997, Business Post, p. 4.
65. Ibid. 8 May 1997, Business Post, p. 1.
66. *USITO China IT Issue Report*, 4.
67. Ibid. 5.
68. Ibid. 6.
69. The State Council General Secretary Luo Gan asserted in a report to the Congress that the restructuring was essential because the existing bureaucracy was too bloated and no longer met economic needs. He also noted that the recent Asian economic crisis posed 'great challenges' to China. *South China Morning Post*, 6 Mar. 1998, p. 11.
70. One government source reportedly noted that Liu's departure was necessary, and quoted a Chinese proverb to say that 'on one mountain there cannot be two tigers', ibid. 7 July 1998, p. 6.
71. Asia Pulse, 27 Apr. 1999 report.
72. *South China Morning Post*, 2 June 1998, Business Post, p. 4.
73. *China Daily*, 2 Dec. 1998.
74. *Financial Times*, 16 Nov. 1998, Survey-China, p. 4.
75. Ibid. 22 Sept. 1998, World Trade, p. 8.

76. Business Times (Singapore), 1 Oct. 1999, East Asia, p. 19.
77. M. Taylor Fravel, 'The Waiting Game: China Unicom', MS written for *Business China*. See http://www.stanford.edu/~fravel/pubs/waitinggame.htm, 28 Feb. 2000.
78. China Online, 3 Sept. 1999 report.
79. *Financial Times*, Telenews Asia Newsletter, 7 Oct. 1999.
80. *China Economic Review*, 28 Sept. 1999, p. 7; *South China Morning Post*, 3 Sept. 1999, Business Post, p. 2.
81. *South China Morning Post*, 2 Feb. 2000, Business Post, p. 3.
82. China Online, 20 Dec. 1999 report.
83. Fravel, 'The Waiting Game'.
84. TIW website. See http://www.tiw.ca/engl/Section1_Infos/C_Medias/year2000/may11.shtml.
85. China Online, 3 Sept. 1999 report.
86. *South China Morning Post*, 17 June 2000, Business Post, p. 1.
87. Ibid. 24 May 1999, p. 3 and 26 May 1999.
88. China Economic Review, Financial Times Information, 23 Sept. 1999.
89. Asiainfo Daily China News, 19 Feb. 2001.
90. China Online, 23 July 2001.
91. Author's observation in Beijing and Shanghai, summer 2002.
92. *South China Morning Post*, 19 May 1999, Business Post, p. 8.
93. Ibid. 10 Jan. 2004, Business Post, p. 2.
94. Ibid. 8 Apr. 1998, Business Post, p. 4.
95. Financial Times Information, Business Daily Update, 11 Jan. 2002.
96. China Online, 8 Feb. 2002.
97. *South China Morning Post*, 25 Mar. 2005, Business Post, p. 1.
98. AFX News Ltd., 20 Jan. 2006 report.
99. Xinhua Economic News Service, 6 July 2007 report.
100. Ibid. 22 May 2000 report.
101. Financial Times Information, International Market Insight Reports, 17 May 2000.
102. China Online, 24 May 2001 report.
103. Comtex News Network, 9 July 2004 report.
104. Financial Times Information, Global News Wire, 1 Apr. 2002.
105. Kyodo News Service, 9 Jan. 1997 report.
106. *South China Morning Post*, 24 May 2000, Business Post, p. 4.
107. Xinhua General News Service, 22 Dec. 2000 report.
108. *China Daily*, 21 June 2000; Xinhua Economic News Service, 17 Oct. 2003; Xinhua Financial News Network, 12 July 2005 reports.
109. MFC Insight, *Carrier Analysis: China Telecom*, 43.
110. Xinhua News Agency, 26 May 1999 report. Building a telecommunications network along a railroad's right of way was not an idea unique to China; in the United States, the Southern Pacific Railroad Company had begun its own

services in 1973 to compete with AT&T. See Susan McMaster, *The Telecommunications Industry*, Westport, Conn.: Greenwood Press, 2002: 105. The railroad company's network later was incorporated into another major telecommunications provide, GTE Sprint.
111. *South China Morning Post*, 27 May 1999, p. 10.
112. Ibid. 15 Aug. 2000, Business Post p. 3; 15 Jan. 2001, Business Post, p. 4.
113. Ibid. 15 Jan. 2001, Business Post, p. 4; 27 May 1999, p. 10.
114. Ibid. 14 May 2001, Business Post, p. 4; 31 Jan. 2004, Business Post, p. 2.
115. *Telecommunications Development Asia–Pacific* website. See http://www.tdap.co.uk/uk/archive/interviews/inter(railcom_0203).html.
116. Comtex News Network, Sinocast report, 8 Dec. 2003 report.
117. Xinhua Economic News Service, 23 Feb. 2005 report. The SASAC replaced the former State Economic and Trade Commission (SETC) in mid-2003.
118. Ibid. July 26, 2004.
119. Comtex News Network, Sinocast report, 28 Sept. 2005.
120. Ibid. 6 Dec. 2005; Xinhua Economic News Service, 23 Feb. 2005 report.
121. From Chinanex report. See http://www.chinanex.com/company/chinasat.htm.
122. AFX Asia report, 9 June 2004.
123. Financial Times Information, Sinocast China, 24 June 2004 report.
124. Xinhua Financial News Network, 4 Nov. 2004 report.
125. Xinhua General News Service, 2 Nov. 2002 report.
126. Ibid. 2 Dec. 2002 report.
127. Xinhua Financial News, 3 Sept. 2003 report.
128. Financial Times Information, 16 July 2003 report.
129. Xinhua Economic News Service, 11 Apr. 2003 report.
130. United Press International, Press Report, 8 Mar. 2002. Further details of the WTO agreement with China are available in 'Report of the Working Party on the Accession of China'.
131. SinoCast China IT Watch report, 15 Dec. 2006. The South Korean company SK Telecom had a joint venture with Unicom, but this was to provide value-added services and content, rather than operate a network. Similarly, Australia's Telstra and other foreign communications companies worked to provide value-added services with Chinese partners. See Daniel Roseman, 'The WTO and Telecommunications Services in China: Three Years On', *Info* 7 (2005): 36.
132. Roseman, 'The WTO and Telecommunications Services in China', 33.
133. This argument was made in Bing Zhang, 'Understanding China's Telecommunications Policymaking and Reforms', *Telematics and Informatics* 19 (2002): 346.
134. Associated Press report, 9 Jan. 2006.
135. Financial Times Information, Global News Wire, 5 July 2006; *Financial Times* (London), 21 Aug. 2006, Companies International, 20.
136. Financial Times Information, Global News Wire, 28 Apr. 2007.
137. *Financial Times* (London), 21 Aug. 2006, Companies International, 20.

138. China Internet Network Information Center (CNNIC), 'Statistical Survey Report on the Internet Development in China', Jan. 2007, p. 5.
139. From Aug. 2004 interview with Datang in Beijing.
140. From 2006 interview with foreign telecommunications analyst in Beijing, 2006; Xinhua Economic News Service, 6 July 2007 report.
141. Xinhua Economic News Service, 6 July 2007 report.
142. Wang had replaced Yang Xianzu as president of Unicom in Aug. 2003.
143. Xinhua Financial Network News, 1 Nov. 2004.

Chapter 4

1. China Internet Network Information Center (CNNIC), 'Statistical Survey Report on the Internet Development in China,' July 2007 report, p. 9, Jan. 2007 report, pp. 5, 6, and 8, and July 1998 report, p. 2.
2. For a comprehensive history of the American data network, see Janet Abbate, *Inventing the Internet*, Cambridge, Mass.: MIT Press, 1999.
3. Ibid. 52, 56.
4. Ibid. 135.
5. Ibid. 191–5.
6. Ibid. 198.
7. Ibid. 197–8.
8. Ibid. 199, 239.
9. Ibid. 203.
10. For a complete history of France's data network, see chs. 4 and 5 in Gunnar Trumbull, *Silicon and the State: French Innovation Policy in the Internet Age*, Washington, DC: Brookings Institution Press, 2004.
11. Ibid. 68–9.
12. The history of Japan's Internet growth is assessed in Ken Coates and Carin Holroyd, *Japan and the Internet Revolution*, New York: Palgrave Macmillan, 2003: ch. 2.
13. Ibid. 46–8.
14. FT Asia Intelligence Wire, Asian Review of Business and Technology, 22 Mar. 2000 report.
15. Ibid. (for 1999 number); Associated Press, 7 Dec. 2005 report (for 2005 number).
16. Some of the following history is derived from Milton Mueller and Zixiang Tan, *China in the Information Age*, Westport, Conn.: Praeger Publishers, 1997: 81–91; Duncan Clark, Alexandra Rehak, and Ted Dean, *The Internet in China*, Beijing: BDA (China) Ltd., 1999: 61–2; Pyramid Research, *Telecommunications Markets in China*, Cambridge, Mass.: The Economist Intelligence Unit, 1999: 123.
17. Clark, Rehak, and Dean, *The Internet in China*, 60.
18. Interview with Chinese Academy of Social Sciences researcher, Beijing, 19 May 1999.
19. Clark, Rehak, and Dean, *The Internet in China*, 56.

20. Zixiang Tan, 'Internet in China', *Pacific Telecommunications Conference Proceedings* (1996): 624.
21. Clark, Rehak, and Dean, *The Internet in China*, 55–6.
22. Ibid. 57.
23. Jonah Greenberg, 'China Netcom: The Little Telecom That Could', Virtual China homepage (7 Dec. 1999). See http://virtualchina.com/infotech/news/stories/120799-netcom.html.
24. Brian Low, 'The Evolution of China's Telecommunications Equipment Market: A Contextual, Analytical Framework', *Journal of Business and Industrial Marketing* 20 (2005): 102; Daniel Roseman, 'The WTO and Telecommunications Services in China: Three Years On', *Info* 7 (2005): 35.
25. For 1994 and 1996: Tan, 'Regulating China's Internet: Convergence Toward a Coherent Regulatory Regime', *Telecommunications Policy* 23 (1999): 263. For 1998–2005, see China Internet Network Information Center (CNNIC), 'Statistical Survey Report on the Internet Development in China', var. year reports. Interactive Audience Measurement Asia (IAMASIA), a Hong Kong-based survey organization reported a lower figure of some 12 million users (defined as someone who has used the Internet within the past four weeks) in mid-2000, when the CNNIC found nearly 18 million users. Note the official statistics are on the same order of magnitude. See the IAMASIA homepage, and China Internet Network Information Center (CNNIC), 'Statistical Survey Report on the Internet Development in China', July 2000 report.
26. Clark, Rehak, and Dean, *The Internet in China*, 96.
27. Sinocast China IT report, 17 Nov. 2004, and *South China Morning Post*, 2 May 2000, Business Post, p. 3.
28. Clark, Rehak, and Dean, *The Internet in China*, 97.
29. Ibid. 155.
30. Paul Triolo and Peter Lovelock, 'Up, Up, and Away—With Strings Attached', *China Business Review* 23 (1996): 29.
31. *Los Angeles Times*, 13 Feb. 1999, p. A2.
32. Interview with telecommunications consultant in Beijing, 2000.
33. China Online homepage, 15 July 1999.
34. *China Daily*, 4 Feb. 1999, online report (see http://www.chinadaily.com.cn/cndydb/1999/02/d5-4net.b04.html/); ibid. 1 Mar. 1999, online report (see http://www.chinadaily.com.cn/cndydb/1999/03/d1-1post.c01.html/); ibid. 2 Mar. 1999, online report (see http://www.chinadaily.com.cn/cndydb/1999/03/d2-5inte.c02.html/); Clark, Rehak, and Dean, *The Internet in China*, 151.
35. Clark, Rehak, and Dean, *The Internet in China*, 137, 151, 102.
36. BDA (China) Ltd., *Broadband Access in China: Focus Report*. Beijing: BDA (China) Ltd., 2000: 35.
37. Sinocast China IT report, 17 Nov. 2004.
38. Financial Times Information, Global News Wire, 28 Mar. 2007 report; and China Internet Network Information Center (CNNIC), 'Statistical Survey

Report on the Internet Development in China', Jan. 2007 report, p. 5. The 91 million figure includes cable modem as well as DSL (digital subscriber line) and other broadband data services.
39. From Sept. 2003 discussion by author with Chinese government official in New York state.
40. From 2006 interview with Beijing-based foreign business person. This is the rate charged his company, but is likely representative of prevalent fees at the time.
41. *Financial Times* (London), Comment and Analysis, p. 21.
42. Financial Times Information, Global News Wire, 19 June 2007 report.
43. *South China Morning Post*, 26 Feb. 1996, p. 8.
44. United Press International, 6 Jan. 1997 report.
45. Associated Press Report, 31 Dec. 1997.
46. *Business Wire* report, 28 Sept. 2000.
47. Agence France Presse, 3 Oct. 2000.
48. Xinhua News Agency, 4 Aug. 2000 report. In Feb. 2001 the Ministry of Public Security released details of its own equivalent of 'NetNanny' filtering software. From Apr. 2001 interview with foreign telecommunications consultant, Beijing.
49. *South China Morning Post*, 9 Feb. 2001. See http://china.scmp.com/technology/ZZZV5PKVPGC.html/.
50. China Internet Network Information Center (CNNIC), 'Statistical Survey Report on the Internet Development in China', var. years.
51. Agence France Presse, 12 Feb. 1999; Deutsche Presse-Agentur, 27 Sept. 2000.
52. Deutsche Presse-Agentur, 27 Sept. 2000.
53. China Internet Network Information Center (CNNIC), 'Statistical Survey Report on the Internet Development in China', Jan. 2007 report.
54. Financial Times Information, Global News Wire, 8 June 2007.
55. Ibid.
56. Deutsche Presse-Agentur, 14 Apr. 2007 report.
57. *Washington Post*, 8 Apr. 1996, p. A01.
58. *Christian Science Monitor*, 21 Mar. 2000, p. 7.
59. *South China Morning Post*, 19 May 2003, p. 4.
60. Human Rights Watch website report. See http://www.hrw.org/advocacy/internet/dissidents/7.htm.
61. *Christian Science Monitor*, 9 Sept. 2005, World, p. 1.
62. United Press International press report, 22 Mar. 2007.
63. *Financial Times* (London), 23 May 2007, World News, p. 10.
64. *South China Morning Post*, 21 Mar. 2000, Technology Post section, p. 4.
65. *New York Times*, 4 Dec. 2002, section A, col. 1, foreign desk, p. 13.
66. Interview with researcher at foreign survey firm, Beijing, June 2000. The company placed software on Internet users' computers, and provided them a small stipend and a guarantee of anonymity for their cooperation with the survey project.
67. *New York Times*, 4 Dec. 2002, sect. A, col. 1, foreign desk, p. 13.

68. Based on personal interviews with several randomly selected taxi drivers in Beijing, July 2006. The compensation issue was driven by a mandated increase in taxi fares in Beijing in early 2006. The subsequent drop in passengers led to hardship for drivers, and the subsidies were meant to alleviate the shortfall for the drivers.
69. Joseph Man Chan, 'Media Internationalization in China', 73.
70. *New York Times*, 25 Jan. 2006, sect. C, col. 5, business-financial desk, p. 3.
71. For a lengthy speech by Wu on the growth of the information industry, see *Renmin youdian* (People's Posts and Telecommunications), 26 Sept. 1998, p. 1. See also *New York Times*, 6 Dec. 2000, section W, p. 1.
72. China Internet Network Information Center (CNNIC), 'Statistical Survey Report on the Internet Development in China', Jan. 2007 report, p. 6.
73. Interview with Shanghai academic researcher, Shanghai, 19 July 2000.
74. *Sydney Morning Herald*, 25 July 2000, p. 33.
75. China Online, 19 July 2000.
76. *Businessweek* Online, 25 Feb. 2005.
77. Ibid. 22 Aug. 2005, pp. 31–2.
78. *South China Morning Post*, 19 Mar. 2002, Business Post, p. 16.
79. Financial Times Information, SinoCast China IT Watch, 14 July 2003 report.
80. Asia Pulse, 21 Dec. 2005 report.
81. *New York Times*, 11 Aug. 2005, section C, international business, p. 4.
82. *New York Times*, 26 June 2007, p. 1.
83. Chan, 'Media Internationalization in China', pp. 79, 76.
84. AFX News Ltd., 20 June 2007 report.
85. China Internet Network Information Center (CNNIC), 'Statistical Survey Report on the Internet Development in China', July 2000 report and July 2003 report, p. 10.
86. China Internet Network Information Center (CNNIC), 'Statistical Survey Report on the Internet Development in China', Jan. 2006 report, p. 61.
87. Norman Nie, and D. Sunshine Hillygus, 'Where Does Internet Time Come From? A Reconnaissance.' *IT&Society* 1 (2002): 14; eid., 'The Impact of Internet Use on Sociability: Time-Diary Findings', ibid. 11.
88. Agence France Presse, 6 Mar. 2007.
89. Guobin Yang, 'Environmental NGOs and Institutional Dynamics in China', *China Quarterly* 181 (2005): 59.
90. *New York Times*, 16 Sept. 2001, sect. 1, p. 20.
91. Informal discussion with Chinese central government official, Apr. 2001, Beijing. This official commented that others in the official's ministry used chat sentiment as a kind of unofficial public opinion poll. Of course, as noted above, such official attention to chat groups left them open to manipulation by overseas Chinese seeking to distort chat discussion content.
92. Reuters World Report, 15 Mar. 2001.
93. *South China Morning Post*, 11 Apr. 2005, Behind the News, p. 18.

94. *New York Times*, 25 Apr. 2005, sect. A, p. 1.
95. Andrew Walder, *Communist Neo-Traditionalism*, Berkeley, Calif.: University of California Press, 1986; and Margaret Pearson, *China's New Business Elite: The Political Consequences of Economic Reform*. Berkeley, Calif.: University of California Press, 1997. Pearson examined private business people and joint venture employees in the early 1990s to reach these conclusions.

Chapter 5

1. Leang-li T'ang, ed. *Reconstruction in China*, Shanghai: China United Press, 1935: 279–80.
2. *Dangdai zhongguo de youdian shiye* (Contemporary China's Posts and Telecommunications Facilities). Beijing: Dangdai zhongguo chubanshe, 1993: 13.
3. Detailed history of the MPT is available in *Zhongguo youdian baike quanshu* (Encyclopedia of China's Posts and Telecommunications), Beijing: Renmin youdian chubanshe, 1995; and in *Dangdai zhongguo de youdian shiye* (Contemporary China's Posts and Telecommunications Facilities).
4. For more information on the various machine building ministries of the 1950s and 1960s, see Chu-Yuan Cheng, *The Machine Building Industry in China*, Aldine-Atherton: Chicago, 1971.
5. Cheng, *Machine Building Industry in China*, 16, 307.
6. *Zhongguo jiaotong nianjian* (China Communications Yearbook), Beijing: Zhongguo jiaotong nianjian she, 1986: 235.
7. *Dangdai zhongguo de youdian shiye* (Contemporary China's Posts and Telecommunications Facilities), 471.
8. Ibid. 456.
9. Information on the early planning system of telecommunications manufacturing is derived from *Dangdai zhongguo de youdian shiye* (Contemporary China's Posts and Telecommunications Facilities), 514–19.
10. *Zhongguo tongji nianjian* (China Statistical Yearbook), 1981: 291.
11. *Zhongguo youdian baike quanshu* (Encyclopedia of China's Posts and Telecommunications). Beijing: Renmin youdian chubanshe, 1995: 485.
12. Ibid. 483.
13. Ibid. 484.
14. Interview with PTIC official in Beijing, 6 July 1995.
15. On profit retention, see Barry Naughton, *Growing Out of the Plan: Chinese Economic Reform, 1978–1993*, New York: Cambridge University Press, 1995: 101–3.
16. Wen was the MPT minister from 1981 to 1984.
17. Xinhua General Overseas News Service, 22 Feb. 1982 report.
18. Ibid. 13 Nov. 1984 report.
19. Ibid. 20 Aug. 1984 report; Tinglong Tang, 'The Development of the Shanghai Local Telephone Service', *China Telecommunications Construction* 1 (1989): 45–6.

20. Dongsheng Cheng and Lili Liu, *Huawei Zhenxiang* (The Truth of Huawei), Beijing: Dangdai zhongguo chubanshe, 2004: 26.
21. Alex Tan, 'The Outlook for the Future of China's CO Switch Market', *Telecom Asia* 7 (Jan. 1996): 41.
22. From 14 July 1995 interview with NEC official in Beijing; Ante Xu and Phillip Armstrong, *Chinese Telecom Market*, New York: Northern Business Information, 1995: 90.
23. Xinhua General Overseas News Service, 15 Aug. 1984 report.
24. Ibid. 4 Sept. 1987 report.
25. Cheng and Liu, *Huawei Zhenxiang* (The Truth of Huawei), 31.
26. Margaret Pearson, *Joint Ventures in the People's Republic of China*, Princeton, NJ: Princeton University Press, 1991.
27. Huasheng Zhou and Maurice Kerkhofs, 'System 12 Technology Transfer to the People's Republic of China', *International Journal of Technology Management* 3 (1988): 205.
28. From July 2002 interview with Alcatel Shanghai Bell, Shanghai.
29. *Business China*, Economist Intelligence Unit, 30 Sept. 1996, p. 8.
30. Xiaobai Shen, *The Chinese Road to High Technology*, New York: St. Martin's Press, 1999: 73.
31. Zhou and Kerkhofs, 'System 12 Technology Transfer', 206, 209.
32. Xinhua General Overseas News Service, 15 July 1991.
33. See e.g. ibid. 25 Apr. 1990 and 22 Aug. 1991.
34. *Beijing Review*, 24–30 June 1996, p. 22.
35. *South China Morning Post*, 28 Aug. 1997, p. 4.
36. Ibid. 25 July 1993, p. 7; *Christian Science Monitor*, 24 Sept. 1991, p. 9.
37. Japan also used this strategy in the 1920s and 30s, when it bought sophisticated switching equipment from various nations so they would keep prices competitive. See Marie Anchordoguy, 'Nippon Telegraph and Telephone Company (NTT) and the Building of a Telecommunications Industry in Japan', *Business History Review* 75 (2001): 513.
38. Zhou and Kerkhofs, 'System 12 Technology Transfer', 211.
39. *South China Morning Post*, 8 Mar. 1994, Business p. 4.
40. Xinhua General Overseas News Service, 22 Aug. 1991.
41. Associated Press, 29 July 1993 report.
42. BBC Summary of World Broadcasts, 2 Sept. 1992 and 7 Oct. 1992 reports.
43. *Asia Pulse*, 9 Apr. 1999 report.
44. *Business China*, Economist Intelligence Unit, 30 Sept. 1996, p. 8.
45. Xing Fan, *China Telecommunications: Constituencies and Challenges*, Cambridge: Program on Information Resources Policy, Center for Information Policy, Harvard University, 1996: 144.
46. This early history of Julong is derived from 'The Great Dragon Group and the HJD-04: Domestic Switching in the Ascendant', *Telecom Asia* 7 (1996): 42–5.

47. The eight companies included the PTIC's Luoyang, Changchun, Hangzhou, and Chongqing factories, MEI factories in Beijing (the Wire Communications Plant), Shenzhen, and Zhengzhou, and one factory owned by the PLA (ibid. 44).
48. Xinhua News Agency, 21 Sept. 1995 report.
49. *Asia Pacific Telecoms Analyst*, 25 Sept. 1995 report.
50. Ibid.
51. Wu is quoted in the newspaper *Jingji Cankao Bao* (Economic Information News), 11 Feb. 1999, cited in China Online report, 16 Feb. 1999.
52. George J. Gilboy, 'The Myth Behind China's Miracle', *Foreign Affairs* 83 (2004): 44–5.
53. From ZTE corporate website. See http://www.zte.com.cn/English/01about/index2.jsp.
54. Ibid.
55. From 23 July 2004 interview with ZTE official in Shenzhen, and Shenzhen Data Communications Bureau website. See http://www.shenzhenwindow.net/topten/e-zte.htm.
56. Xinhua News Agency 19 June 1996, report and *Asia Pulse*, 9 Apr. 1999 report.
57. From 23 July 2004 interview with ZTE official, Shenzhen.
58. Ibid.
59. From Shenzhen Data Communications Bureau website. See http://www.shenzhenwindow.net/topten/e-zte.htm.
60. *Financial Times*, 5 Nov. 1998, World Trade, p. 11.
61. Ibid.
62. Asia Pulse, 9 Apr. 1999 report.
63. Li Sun, 'Huawei: Tulang Xiang Shizi de Yanjiang' (The Evolution from a Wolf to a Lion), *IT Jingli Shijie* (CEO and CIO Information Times), 10 (2002): 50–62; Cheng and Liu, *Huawei Zhenxiang*, 26.
64. Forbes online. See http://www.forbes.com/global/2002/1111/056sidebar.html.
65. As early as 1993, when private firms in socialist China were just establishing a beachhead on the economic front, the official government news agency Xinhua referred to Huawei as a 'non-governmental technology research and development enterprise'. From Xinhua General Overseas News Service, 3 Feb. 1993 report.
66. Cheng and Liu, *Huawei Zhenxiang*, 26–7. The following discussion of Huawei's early history, unless otherwise noted, is derived from this source's ch. 1.
67. Xinhua News Agency, 15 Nov. 1994 report.
68. *Financial Times*, 11 Jan. 2005, comment and analysis, p. 15.
69. Cheng and Liu, *Huawei Zhenxiang*, 30 [author's translation].
70. *Wall Street Journal*, 28 Aug. 1997, p. A11.
71. Electronic Buyers News, 24 Feb. 1997 report.
72. Asia Pulse, 9 Apr. 1999 report.
73. Sun, 'Huawei: Tulang Xiang Shizi de Yanjiang' (The Evolution from a Wolf to a Lion), 55.
74. Xinhua News Agency, 9 Apr. 1997 report.
75. Financial Times Information, Newsbase Russian Daily Bulletin, 21 Oct. 1999 report.

76. Cheng and Liu, *Huawei Zhenxiang*, 105–6. The following paragraph is based on assertions in their work and in the study by Sun cited above, 'Huawei: Tulang Xiang Shizi de Yanjiang' (The Evolution from a Wolf to a Lion).
77. Sun, 'Huawei: Tulang Xiang Shizi de Yanjiang' (The Evolution from a Wolf to a Lion), 53; Cheng and Liu, *Huawei Zhenxiang*, 105–6.
78. Sun, 'Huawei: Tulang Xiang Shizi de Yanjiang' (The Evolution from a Wolf to a Lion), 53–4.
79. Based on interview with foreign telecommunications consultant in Beijing, July 2004, and with Huawei official in Shenzhen, July 2004.
80. United States Foreign and Commercial Service report, entitled 'China-Telecom Operators Urged to Buy Locally-Produced Equipment', cited in Financial Times Asia Intelligence Wire, 12 Jan. 1999.
81. *China Economic Review*, Feb. 1995, p. 8.
82. Xinhua News Agency, 3 June 1998 report.
83. Xinhua General News Service, 19 May 2001; China Online, 20 Dec. 2001 report.
84. From July 2002 interview with Alcatel Shanghai Bell, Shanghai.
85. Ibid.
86. *South China Morning Post*, 12 Aug. 1998, p. 6.
87. Business Daily Update, 6 June 2001.
88. From July 2002 interview with Alcatel Shanghai Bell, Shanghai.
89. *China Daily*, 17 June 2002.
90. *South China Morning Post*, 21 Oct. 2002, Business Post, p. 4; Financial Times Information, SinoCast News Wire, 21 July 2003 report; Aug. 2004 interview with Siemens representative, Beijing.
91. From July 2002 interview with Alcatel Shanghai Bell, Shanghai.
92. AFX News Limited, 23 Oct. 2001 report.
93. From July 2002 interview with Alcatel Shanghai Bell, Shanghai.
94. From August 2004 interview, Beijing.
95. Xinhua Financial News, 15 Jan. 2004 report; and Financial Times Information, SinoCast China IT Watch, 15 Jan. 2004 report.
96. Xinhua Financial News, 15 Jan. 2004 report.
97. Financial Times Information, China Online, 16 Dec. 1999 report.
98. *South China Morning Post*, 5 Apr. 2003, Business Post, p. 2.
99. Financial Times Information, Asia Africa Intelligence Wire, 24 Jan. 2003 report.
100. From Huawei Corporation, *Your Profit, Our Goal*, 2004: 10.
101. AFX News Ltd., 9 June 2005 report.
102. Asia Pulse, 18 Nov. 2004 report.
103. Ibid.; Cheng and Liu, *Huawei Zhenxiang*, 53; July 2004 interview with ZTE official, Shenzhen.
104. *Financial Times*, 11 Jan. 2005, comment and analysis, p. 15.
105. From ChinaNex report, online. See http://www.chinanex.com/company/putian.htm.

106. *Businessweek*, 8 Nov. 2004, p. 77.
107. *Wall Street Journal*, 1 June 2005. See http://online.wsj.com/article/ 0,,SB111756576875747309,00.html?mod=yahoo_hs&ru=yahoo.

Chapter 6

1. Statistics on GDP and urban and rural income from *Zhongguo tongji nianjian* (China Statistical Yearbook), Beijing: Zhongguo tongji chubanshe, 2007: 67 (for 2006 Shanghai GDP); ibid. 2007: 355, 369 (for 2006 income); and *Statistical Yearbook of Shanghai*, Beijing: China Statistics Press, 1996, p. 26 (for 1984 GDP).
2. Xinhua News Service, 30 Dec. 2001 report.
3. For 2006 mobile phones, see: *Shanghai Statistical Yearbook*, 2007: 296. For Internet users and websites, see China Internet Network Information Center (CNNIC), 'Statistical Survey Report on the Internet Development in China', Jan. 2007 report, pp. 19, 20.
4. Stephen Graham, 'Cities, Nations and Communications in the Global Era: Urban Telecommunications Policies in France and Britain', *European Planning Studies* 3 (1995): 367–70.
5. Darrene Hackler, 'Invisible Infrastructure and the City', *American Behavioral Scientist* 46 (2003): 1034–55.
6. Paul Sommers and Daniel Carson, 'What the IT Revolution Means for Regional Economic Development', Washington, DC: Brookings Institution Center on Urban and Metropolitan Policy (2003): 1–54.
7. Susan Walcott and James Wheeler, 'Atlanta in the Telecommunications Age: The Fiber-optic Information Network', *Urban Geography* 22 (2001): 334, cited in Darrene Hackler, 'High-Tech Growth and Telecommunications Infrastructure in Cities', *Urban Affairs Review* 39 (2003): 62.
8. Stephen Graham, 'Global Grids of Glass', *Urban Studies* 36 (1999): 940.
9. Jean Oi, 'Fiscal Reform and the Economic Foundations of Local State Corporatism in China', *World Politics* 45 (1992): 102
10. Ibid. 124.
11. Gang Tian, *Shanghai's Role in the Economic Development of China*, Westport, Conn.: Praeger, 1996: 61; emphasis in the original.
12. Segal and Thun, 'Thinking Globally, Acting Locally: Local Governments, Industrial Sectors, and Development in China', *Politics and Society* 29 (2001): 581 and 583.
13. Eric Harwit, *China's Automobile Industry: Policies, Problems, and Prospects*, Armonk, NY: M. E. Sharpe, 1995.
14. Unless otherwise indicated, the pre-1949 historical information in this section is based on *Today's Shanghai P&T*, Shanghai: Shanghai Posts and Telecommunications Authority, 1994: 34–6. A more detailed discussion of the earliest telecommunications development in imperial China can be found in Baark's *Lightning*

Wires: The Telegraph and China's Technological Modernization, Westport, Conn.: Greenwood Press, 1997.
15. Tinglong Tang, 'The Development of the Shanghai Local Telephone Service', *China Telecommunications Construction* 1 (1989): 43–4.
16. The Beijing representative of Shanghai's largest foreign investor, Volkswagen, often had to wait up to thirty minutes to place a call to the Shanghai office, even as late as 1988. From author's interview in Beijing.
17. *Zhongguo tongji nianjian* (China Statistical Yearbook), 1998: 570.
18. The GDP grew about fourfold on a national scale, and a bit less in Shanghai over these years. Derived from *Zhongguo tongji nianjian* (China Statistical Yearbook), 2003: 55 and *Statistical Yearbook of Shanghai*, 1996: 26.
19. From author's interview with telecommunications official, Shanghai.
20. *Statistical Yearbook of Shanghai*, 1992: 467.
21. Tang, 'The Development of the Shanghai Local Telephone Service', 45–6.
22. Xinhua General Overseas News Service, 15 July 1986.
23. From author's interview with telecommunications official, Shanghai.
24. *Statistical Yearbook of Shanghai*, 1999: 50.
25. Xinhua News Service, 30 Dec. 2001.
26. From author's interview, Beijing.
27. Author's own observation, Shanghai, Aug. 2004.
28. China Internet Network Information Center (CNNIC), 'Statistical Survey Report on the Internet Development in China,' various year reports.
29. *South China Morning Post*, 12 June 2000, Business Post, p. 3.
30. Segal and Thun, 'Thinking Globally, Acting Locally'; Tian, *Shanghai's Role in the Economic Development of China*; Xiaobai Shen, *The Chinese Road to High Technology*, New York: St Martin's Press, 1999.
31. *Financial Times* (London), 6 Nov. 2002, p. 21.
32. China Online, 2 Dec. 1998.
33. Xinhua General News Service, 5 July 2000.
34. *South China Morning Post*, 15 Nov. 2001, Business Post, p. 4.
35. Financial Times Information, Business Daily Update, 27 Oct. 2003 report.
36. From LeadDiscovery website. See http://www.leaddiscovery.com.cn/en/intro/unit1.htm.
37. Asia Pulse (Nationwide Financial News), 26 Feb. 1999.
38. *People's Daily*, 9 Feb. 2000.
39. Two excellent studies of Chinese corruption are those by Julia Kwong, *The Political Economy of Corruption in China*, Armonk, NY: M. E. Sharpe, 1997; Xiaobo Lü, *Cadres and Corruption: The Organizational Involution of the Chinese Communist Party*, Stanford, Calif.: Stanford University Press, 2000
40. Lü, *Cadres and Corruption*, 246.
41. For more information on the early development of China Unicom, see Eric Harwit and Jack Su, 'A Telecom Newcomer Challenges the MPT Monopoly', *China Business Review* 23 (1996): 22–3. Unless otherwise noted, information on

the early development of Shanghai Unicom is based on author's interview with a Shanghai Unicom official in Shanghai.
42. *Asia Pacific Telecoms Analyst*, 9 Oct. 1995, p. 16.
43. For details on Liu's background, see *New York Times*, 18 June 1993, p. D2.
44. *Hong Kong Standard*, 7 Oct. 1997.
45. From 1997 author's interview with Shanghai Unicom official in Shanghai.
46. *South China Morning Post*, 13 July 1995, China Business Review section, p. 1.
47. Ibid. 2 Feb. 2000, Business Post, p. 3.
48. Asiainfo Daily China News, 20 Feb. 2001.
49. Xinhua News Agency, 16 Apr. 2001.
50. For this and more detail on the early history of AT&T's ties with China, see Xing Fan, *China Telecommunications: Constituencies and Challenges*, Cambridge: Program on Information Resources Policy, Center for Information Policy, Harvard University, 1996: 134–40.
51. United Press International, 31 May 1989.
52. Associated Press, 19 Aug. 1993.
53. China Online, 24 Apr. 2002.
54. From author's interview with AT&T-China president Arthur Kobler in Shanghai. The following discussion on AT&T's development in the 1990s is based on this interview.
55. From author's interview in July 2002 with Shanghai Symphony president and general manager Rick Luk. The following discussion, unless otherwise noted, is based on this interview.
56. *Financial Times*, 6 Dec. 2000, p. 35.
57. China Online, 19 Dec. 2001.
58. *China Daily* online report, 28 Mar. 2007.
59. For a further discussion of the issue of Internet censorship, see Michael Chase and James Mulvenon, *You've Got Dissent! Chinese Dissidents Use of the Internet and Beijing's Counter-strategies*. Santa Monica, Calif.: RAND, 2002
60. Lynn T. White, 'All the News: Structure and Politics in Shanghai's Media Reform', in Chin-Chuan Lee, ed., *The Interplay of Politics and Journalism*, New York: The Guilford Press, 1990: 88–110.
61. China Internet Network Information Center (CNNIC), 'Statistical Survey Report on the Internet Development in China', Jan. 2007 report, pp. 6, 20.
62. *South China Morning Post*, 26 Dec. 1995, Business, p. 1; China Online, 21 July 2000.
63. From author's interviews with two Shanghai media researchers in Shanghai; China Online, 21 July 2000.
64. Xinhua News Service, 28 May 2000.
65. China Online, 16 June 2000.
66. The following information is based on author's interview in Shanghai with a Shanghai businessman familiar with Eastday.com's café operations.
67. Former President Jiang Zemin gave up his last official post, as head of the communist party's military commission, in Sept. 2004. This stood to weaken

the hand of his son, Jiang Mianheng, and subsequently affect the power of the city's municipal corporations.
68. Though it is impossible to rule out the factor of corruption in the development of Shanghai's municipal telecommunications system, it is noteworthy that neither my own primary interview material nor secondary accounts of the sector indicate corruption was present in a significant degree in the construction or operation of the voice and data networks.

Chapter 7

1. Heather Hudson, *When Telephones Reach the Village*, Norwood, NJ: Ablex Publishing Corp., 1984.
2. Ibid. 43–4. These assertions are derived from Andrew P. Hardy, 'The Role of the Telephone in Economic Development,' and id., 'The Role of the Telephone in Economic Development: An Empirical Analysis', Geneva: International Telecommunications Union publication, 1981. According to Hudson, Hardy's analysis found that a 1 percent rise in the number of telephones per 100 population in a nine-nation study between 1950 and 1955 contributed to a rise in per capita GDP between 1955 and 1962 of about 3 percent.
3. E.B. Parker, 'Economic and Social Benefits of the REA Telephone Loan Program', Mountainview, Calif.: Equatorial Communications, 1981, cited in Hudson, *When Telephones Reach the Village*, 52.
4. Lars-Hendrick Röller and Leonard Waverman, 'Telecommunications Infrastructure and Economic Development: A Simultaneous Approach', *American Economic Review* 91 (2001): 909–23.
5. Ibid. 921.
6. S. N. Kaul, 'India's Rural Telephone Network', New Delhi: Economics Study Cell, Posts and Telegraphs Boards, Ministry of Communications of India, 1981, cited in Hudson, *When Telephones Reach the Village*, 63.
7. D. G. Clarke and W. Laufenberg, 'The Role of Telecommunications in Economic Development', with Special Reference to Rural Sub-Sahara Africa', Geneva: International Telecommunications Union, 1981, cited in Hudson, *When Telephones Reach the Village*, 70.
8. Paul Mundy and Jacques Sultan, *Information Revolutions*, Wageningen, Netherlands: CTA, 2001: 105.
9. Morten Falch and Amos Anymadu, 'Tele-centres as a Way of Achieving Universal Access—the Case of Ghana', *Telecommunications Policy* 27 (2003): 31.
10. Hudson, *When Telephones Reach the Village*, 57.
11. Hudson's assertion is based on Kaul's study of India, cited above.
12. Cited in Jack Linchuan Qiu, 'Coming to Terms with Informational Stratification in the People's Republic of China', *Cardozo Arts and Entertainment* 20 (2002): 158.
13. Organization for Economic Co-operation and Development (OECD), 'Understanding the Digital Divide', 2001, published online, p. 4.

14. A further way of evaluating the divide would be intra-regional or perhaps intra-municipal evaluation; however, government statistics do not document or tabulate this diversity in any sort of comprehensive national way.
15. Tinglong Tang, 'The Development of the Shanghai Local Telephone Service', *China Telecommunications Construction* 1 (1989): 43–4.
16. The following historical discussion, unless otherwise noted, is derived from *Dangdai zhongguo de youdian shiye* (Contemporary China's Posts and Telecommunications Facilities), Beijing: Dangdai zhongguo chubanshe, 1993: 283–94.
17. The historian Maurice Meisner refers to the inflation of production numbers during this period as the 'wind of exaggeration'. Meisner, *Mao's China and After*, New York: Free Press, 1999: 237.
18. *Dangdai zhongguo de youdian shiye* (Contemporary China's Posts and Telecommunications Facilities), 287.
19. Ibid. 344–5.
20. For detail on these policies, see Xiangdong Wang, *Xinxihua: Zhongguo 21 shiji de xuanze* (Informatization: China's Choices in the 21st century), Beijing: Shehuikexue wenzhai chubanshe, 1998: 360–2 and John Ure, 'Telecommunications in China and the Four Dragons', in id. ed., *Telecommunications in Asia*, Hong Kong: Hong Kong University Press, 1997: 16.
21. *South China Morning Post*, 28 Aug. 1997, China Business Review section, p. 2.
22. In 1995, China had about 859 million rural and 352 million urban residents. From *Zhongguo tongji nianjian* (China Statistical Yearbook), 1998: 105.
23. Ibid. 1995, p. 495; *China Posts and Telecommunications Annual Report*, 1996: 33.
24. For percentages of areas connected, see *Dangdai zhongguo de youdian shiye* (Contemporary China's Posts and Telecommunications Facilities), 289. For numbers of townships and villages, see *Zhongguo tongji nianjian* (China Statistical Yearbook), 2001: 363.
25. *Asia Pacific Telecoms Analyst*, 5 Feb. 1996. The number of villages is based on the number of 'villagers' committees' recorded for the years since 1996, and represents administrative reorganization that reduced the number of villages from the 1985 count. Figure is from *Zhongguo tongji nianjian* (China Statistical Yearbook), 2001: 363.
26. *South China Morning Post*, 11 June 1998, Business Review, p. 1.
27. China Online, 9 Mar. 1999 report.
28. Xinhua Economic News Service, 6 July 2001.
29. *China Daily*, 17 Oct. 1999, p. 5; and Financial Times Information, 13 Feb. 2003.
30. *Zhongguo tongji nianjian* (China Statistical Yearbook), 2006, CD-ROM version, chart 16–43.
31. Ibid. 2001, p. 538.
32. Research Center for Regulation and Competition, 'Universal Service Obligations in China's Telecom Sector', 72 for definition of phone access.
33. Ibid. 72.
34. *Zhongguo tongji nianjian* (China Statistical Yearbook), 1998: 568.

35. Research Center for Regulation and Competition, 'Universal Service Obligations in China's Telecom Sector', 18.
36. For 2001 numbers, see ibid.; for 1999 figures, see *Zhongguo jiaotong nianjian* (China Communications Yearbook), 2000: 667.
37. *Zhongguo tongji nianjian* (China Statistical Yearbook), 2006, CD-ROM version, chart 16–42 for mobile phones; and populations from *Zhongguo tongji nianjian* (China Statistical Yearbook), 2005: 100.
38. *Dangdai zhongguo de youdian shiye* (Contemporary China's Posts and Telecommunications Facilities), 291.
39. Ibid. 292–3. Conversion to dollar numbers use exchange rates for the relevant years.
40. Li Xu, 'Chaju, wenti, duice: dui woguo tongxin qu cheng fazhan bu pingheng de sikao' (Disparities, Problems, and Countermeasures: Thoughts on China's Uneven Regional Communications Development), *Zhongguo ruankexue* (Journal of China Soft Science) (May 1997): 44–51.
41. *Interfax News Agency, China Business*, 8 Mar. 2002.
42. Xinhua Economic News Service, 24 Jan. 2003.
43. Xu, 'Chaju, wenti, duice: dui woguo tongxin qu cheng fazhan bu pingheng de sikao' (Disparities, Problems, and Countermeasures: Thoughts on China's Uneven Regional Communications Development), 51.
44. Huaide Liu and Hanhui Hu, 'Lun chuangxin shangpin xiaofei de baofaxing' (On the Eruptiveness of Consumption of Creative Goods), *Caijing yanjiu* (The Study of Finance and Research), 7 (2001): 15.
45. Angang Hu and Shaojie Zhou, 'Zhongguo de xinxihua zhanlüe: suoxiao xinxi chaju' (China's Informatization Strategy: Reducing the Information Gap), *Guomin jingji guanli* (National Economic Management) (Jan. 2001): 25–9.
46. China Internet Network Information Center (CNNIC), 'The Internet Timeline of China 1987–1996'.
47. Eric Harwit and Duncan Clark, 'Shaping the Internet in China', *Asian Survey* 41 (2001): 388–90; figure of average monthly income for Beijing urban residents is for 1997; from *Zhongguo tongji nianjian* (China Statistical Yearbook), 1998: 332.
48. Harwit and Clark, 'Shaping the Internet in China', 391.
49. Author's observations in Beijing, May 2001.
50. Percentages derived from base number of users in each province over the period 2000 to 2006, from China Internet Network Information Center (CNNIC), 'Statistical Survey Report on the Internet Development in China', Jan. 2001: 6 (number of users derived from percentage noted in the survey); and Jan. 2007 report, p. 19.
51. *China Daily* online, 13 Feb. 2003. See http://www1.chinadaily.com.cn/cndy/2003-02-13/104410.html.
52. Figures on rural and urban population numbers from *Zhongguo tongji nianjian* (China Statistical Yearbook), 2002: 93.
53. China Internet Network Information Center (CNNIC), 'Statistical Survey Report on the Internet Development in China', Jan. 2007 report, p. 5.

54. *China Daily*, 13 Feb. 2003.
55. BBC Worldwide Monitoring, 8 July 2005 report.
56. For a discussion of the problems of rural education in the 1990s and 2000s, see June Teufel Dreyer, *China's Political System: Modernization and Tradition*, New York: Pearson Longman, 2006: 224–7.
57. Figures on literacy from *Zhongguo tongji nianjian* (China Statistical Yearbook), 2007: 120.
58. Tibet's urban income, most likely skewed by the tourism-centered provincial capital of Lhasa, has an atypically high income level. Tibet's rural income level is a better indicator of provincial income levels.
59. The senior middle school education or higher percentages for those over six years of age in these provinces as of the year 2006 were: Xinjiang (20), Guangxi (16), and Ningxia (19), among those with higher Internet penetration. At lower penetration levels, the percentages were: Anhui (14), Guizhou (9), and Tibet (4). Derived from *Zhongguo tongji nianjian* (China Statistical Yearbook), 2007: 118–19.
60. Wei Bu and Linchuan Qiu, 'Report on Media Usage Among Different Groups in Sichuan' (May 2002) (unpub. MS in electronic form).
61. Edwin B. Parker and Heather E. Hudson, *Electronic Byways*, Boulder, Colo.: Westview Press, 1992: 205.
62. From the Universal Service Administrative Company website. See http://www.usac.org/about/universal-service/.
63. Ibid. See http://www.usac.org/about/universal-service/fund-facts/fund-facts-all-program-data.aspx.
64. Personal correspondence with Duncan Clark, managing director, BDA (China) Ltd., June 2002.
65. Ibid.
66. Financial Times Information, 13 Feb. 2003 report.
67. *China Daily*, 27 Apr. 2007, p. 13.
68. Milton Mueller and Zixiang Tan, *China in the Information Age*, Westport, Conn.: Praeger Publishers, 1997: 86.
69. Duncan Clark, Alexandra Rehak, and Ted Dean, *The Internet in China*. Beijing: BDA (China) Ltd., 1999: 61.
70. Interview with CERNET official, Jan. 2003. Unless otherwise noted, the following information on CERNET is based on this interview. At the end of 2002, China had 1,396 institutions of higher education (from *Zhongguo tongji nianjian* (China Statistical Yearbook), 2003: 718).
71. *Business Wire*, 21 Sept. 2005 report.
72. Author's observation of Internet access rules at Beijing University, summer, 2002.
73. Statistics on secondary and primary schools for 2004, from *Zhongguo tongji nianjian* (China Statistical Yearbook), 2005: 691.
74. China Online, 15 Feb. 2001.
75. *Interfax News Agency, China Business*, 11 Jan. 2002.
76. *Business Wire*, 17 Aug. 2001.

77. UNESCO's International Bureau of Education website. See http://www.ibe.unesco.org/International/ICE/46english/46ws6e.htm.
78. The following paras. discussing wireless technology and its relative cost are derived from Research Center for Regulation and Competition, 'Universal Service Obligations in China's Telecom Sector', 72–3.
79. Ibid. 73.
80. Financial Times Information, Global News Wire, 13 Apr. 2006 report; and author's observations of handset prices in Beijing, 2006.
81. Xinhua Financial Network News, 19 June 2007 report.
82. ChinaNex report, 2005. See http://www.chinanex.com/service/adsl.htm.
83. From Sept. 2003 discussion by author with Chinese government official in New York state.
84. Details on one such program are found in a 10 Apr. 2003 report by CNN. See http://www.cnn.com/2003/TECH/ptech/04/10/highspeed.house.ap/index.html.

Chapter 8

1. The following discussion is derived from Peter Evans, *Embedded Autonomy: States and Industrial Transformation*, Princeton, NJ: Princeton University Press, 1995: 235–40.
2. On Bangalore, see e.g. Prashanth Bharadway, 'Bangalore, the Silicon Valley of India', *Competitiveness Review* 15 (2005) (2 pp. preceding p. 82).
3. *The Economist*, 3 Mar. 2005, online report. See http://www.economist.com/surveys/displaystory.cfm?story_id=3689274.
4. *South China Morning Post*, 18 May 2006, news, p. 6.

Bibliography

Books and Articles

Abbate, Janet. *Inventing the Internet*. Cambridge, Mass.: MIT Press, 1999.

Akwule, Raymond U. 'Telecommunciations in Africa: Policy and Management Trends'. In Meheroo Jusawalla, ed. *Global Telecommunications Policies*. Westport, Conn.: Greenwood Press, 1993: 159–70.

Anchordoguy, Marie. 'Nippon Telegraph and Telephone Company (NTT) and the Building of a Telecommunications Industry in Japan'. *Business History Review* 75 (2001): 507–41.

Baark, Erik. *Lightning Wires: The Telegraph and China's Technological Modernization*. Westport, Conn.: Greenwood Press, 1997.

Barnett, A. Doak. *Cadres, Bureaucracy, and Political Power in Communist China*. New York: Columbia University Press, 1976.

Bartke, Wolfgang. *Who was Who in the People's Republic of China*. Munich: K. G. Saur Verlag, 1997.

BDA (China) Ltd. *Broadband Access in China: Focus Report*. Beijing: BDA (China) Ltd., 2000.

BDAssociates Ltd. *China Telecom Update*. Hong Kong, Apr. 1996.

Bharadway, Prashanth. 'Bangalore, the Silicon Valley of India'. *Competitiveness Review* 15 (2005) (2 pp. preceding p. 82).

Brooks, John. *Telephone: The First Hundred Years*. New York: Harper & Row, 1975.

Bu, Wei, and Qiu, Linchuan. 'Report on Media Usage Among Different Groups in Sichuan'. (May 2002) (unpub. MS in electronic form).

Chan, Joseph Man. 'Media Internationalization in China: Processes and Tensions'. *Journal of Communication* 44 (Sept. 1994): 70–88.

Chase, Michael, and Mulvenon, James. *You've Got Dissent! Chinese Dissidents Use of the Internet and Beijing's Counter-strategies*. Santa Monica, Calif.: RAND, 2002.

Cheng, Chu-Yuan. *The Machine Building Industry in China*. Chicago: Aldine-Atherton, 1971.

Cheng, Dongsheng, and Liu, Lili. *Huawei Zhenxiang* (The Truth of Huawei). Beijing: Dangdai zhongguo chubanshe, 2004.

China Internet Network Information Center (CNNIC). 'The Internet Timeline of China 1987–1996'. Beijing: China Internet Network Information Center (2003). Online at http://www.cnnic.net.cn/html/Dir/2003/12/12/2000.htm.

Bibliography

China Internet Network Information Center (CNNIC). 'Statistical Survey Report on the Internet Development in China'. Beijing: China Internet Network Information Center, var. years. Online at http://cnnic.com.cn/en/index/0O/02/index.htm.

China Posts and Telecommunications Annual Report. Beijing: China Posts and Telecommunications Editorial Group of MPT, various years.

Clark, Duncan, Rehak, Alexandra, and Dean, Ted. *The Internet in China.* Beijing: BDA (China) Ltd., 1999.

Clarke, D. G., and Laufenberg, W. 'The Role of Telecommunications in Economic Development, with Special Reference to Rural Sub-Sahara Africa'. Geneva: International Telecommunications Union, 1981.

Coates, Ken, and Holroyd, Carin. *Japan and the Internet Revolution.* New York: Palgrave Macmillan, 2003.

Cortell, Andrew P. *Mediating Globalization.* Albany, NY: State University of New York Press, 2006.

Craig, Jack. 'China: Domestic and International Telecommunications, 1949-1974'. In Joint Economic Committee, Congress of the United States, *China: A Reassessment of the Economy.* Washington, DC: US Government Printing Office, 1975: 289-310.

Crawcour, E. Sydney. 'Industrialization and Technological Change, 1885-1920'. In Kozo, Yamamura ed. *The Economic Emergence of Modern Japan.* Cambridge: Cambridge University Press, 1997: 50-115.

Dangdai zhongguo de youdian shiye (Contemporary China's Posts and Telecommunications Facilities). Beijing: Dangdai zhongguo chubanshe, 1993.

Donnithorne , Audrey. *China's Economic System.* New York: Praeger, 1967.

Doran, Clare, and Stansell, Christina M. 'Nippon Telegraph and Telephone Corporation'. In Tina Grant, ed. *International Directory of Company Histories* 51. Detroit, Mich.: St James Press, 2003: 271-5.

Dreyer, June Teufel. *China's Political System: Modernization and Tradition.* New York: Pearson Longman, 2006.

Ehrlich, Éva. *Japan: A Case of Catching Up.* Budapest: Akadémiai Kiadó, 1984.

Ellis, Clyde T. *A Giant Step.* New York: Random House, 1966.

Evans, Peter. 'Class, State, and Dependence in East Asia: Lessons for Latin Americanists.' In Frederic C. Deyo, ed. *The Political Economy of the New Asian Industrialism.* Ithaca, NY: Cornell University Press, 1987: 203-26.

—— *Embedded Autonomy: States and Industrial Transformation.* Princeton, NJ: Princeton University Press, 1995.

Falch, Morten, and Anymadu, Amos. 'Tele-centres as a Way of Achieving Universal Access—the Case of Ghana'. *Telecommunications Policy* 27 (2003): 21-39.

Fan, Xing. *China Telecommunications: Constituencies and Challenges.* Cambridge: Program on Information Resources Policy, Center for Information Policy, Harvard University, 1996.

Feuerwerker, Albert. *China's Early Industrialization.* Cambridge, Mass.: Harvard University Press, 1958.

Bibliography

Fravel, M. Taylor. 'The Waiting Game: China Unicom'. MS written for *Business China* and posted at http://www.stanford.edu/~fravel/pubs/waitinggame.htm, 28 Feb. 2000.

Gilboy, George J. 'The Myth Behind China's Miracle'. *Foreign Affairs* 83 (2004): 33–48.

Godemont, François. *The Downsizing of Asia*. New York: Routledge, 1999.

Gong, Dacai, ed. 'Youdian tongxinye' (Post and Telecommunications Industry). In Hong Ma, and Shangqing Sun, eds. *Xiandai zhongguo jingji da shi dian* (Compendium on Modern China's Economy). Beijing: Zhongguo caizheng jingji chubanshe, 1993: 1844–6.

Graham, Stephen. 'Cities, Nations and Communications in the Global Era: Urban Telecommunications Policies in France and Britain'. *European Planning Studies* 3 (1995): 357–80.

—— 'Global Grids of Glass'. *Urban Studies* 36 (1999): 929–49.

'The Great Dragon Group and the HJD-04: Domestic Switching in the Ascendant'. *Telecom Asia* 7 (1996): 42–5.

Greenberg, Jonah. 'China Netcom: The Little Telecom That Could'. Virtual China homepage (7 Dec. 1999). Online at http://virtualchina.com/infotech/news/stories/120799-netcom.html.

'Guanyu Jiejue Liantong Gongsi Fazhan Wenti de Diaocha Baogao' (A survey findings report on solving the Unicom company's development problems). Internal memo from the State Economic and Trade, Planning, and State System Restructuring Commissions to then-vice premier Zou Jiahua and then-vice premier Wu Bangguo, 14 Mar. 1997.

Hackler, Darrene. 'Invisible Infrastructure and the City'. *American Behavioral Scientist* 46 (2003): 1034–55.

—— 'High-Tech Growth and Telecommunications Infrastructure in Cities'. *Urban Affairs Review* 39 (2003): 59–86.

Haggard, Stephan, and Cheng, Tun-jen. 'State and Foreign Capital in the East Asian NICs'. In Frederic C. Deyo, ed. *The Political Economy of the New Asian Industrialism*, Ithaca, NY: Cornell University Press, 1987: 84–129.

Hanson, Jarice, and Narula, Uma. *New Communication Technologies in Developing Countries*. Hillsdale, NJ: Lawrence Erlbaum Associates, 1990.

Hardy, Andrew P. 'The Role of the Telephone in Economic Development'. *Telecommunications Policy* 4 (1980): 278–86.

—— 'The Role of the Telephone in Economic Development: An Empirical Analysis'. Geneva: International Telecommunications Union publication, 1981.

Harwit, Eric. *China's Automobile Industry: Policies, Problems, and Prospects*. Armonk, NY: M. E. Sharpe, 1995.

—— 'China's Telecommunications Industry: Development Patterns in Policies'. *Pacific Affairs* 71 (1998): 175–93.

—— and Clark, Duncan. 'Shaping the Internet in China'. *Asian Survey* 41 (2001): 377–408.

Bibliography

Harwit, Eric. and Su, Jack. 'A Telecom Newcomer Challenges the MPT Monopoly'. *China Business Review* 23 (1996): 22–3.

Harrington, Andrew. 'Companies and Capital in Asia–Pacific Telecommunications'. In John Ure, ed. *Telecommunications in Asia*. Hong Kong: Hong Kong University Press, 1997: 81–110.

Horizon House Publications, Inc. 'China: Towards the World's Largest Market'. *Telecommunications International Edition* 30 (1996): S1–S34.

Hu, Angang, and Zhou, Shaojie. 'Zhongguo de xinxihua zhanlüe: suoxiao xinxi chaju' (China's Informatization Strategy: Reducing the Information Gap). *Guomin jingji guanli* (National Economic Management) (Jan. 2001): 25–9.

Huawei Corporation, *Your Profit, Our Goal*. 2004.

Hudson, Heather. *When Telephones Reach the Village*. Norwood, NJ: Ablex Publishing Corp., 1984.

Johnson, Chalmers. *MITI and the Japanese Miracle*. Stanford, Calif.: Stanford University Press, 1982.

Kalathil, Shanthi, and Boas, Taylor. 'Wired for Modernization in China'. In Shanthi Kalathil, and Taylor Boas, *Open Networks, Closed Regimes: The Impact of the Internet on Authoritarian Rule*. Carnegie Endowment for International Peace, Jan. 2003: 13–42. Online at http://www.ciaonet.org/book/kas01/kas01b.pdf.

Katz, Richard. *Japan, the System that Soured: The Rise and Fall of the Japanese Economic Miracle*. Armonk, NY: M. E. Sharpe, 1998.

Kaul, S. N. 'India's Rural Telephone Network'. New Delhi: Economics Study Cell, Posts and Telegraphs Boards, Ministry of Communications of India, 1981.

Krugman, Paul. 'The Myth of Asia's Miracle'. *Foreign Affairs* 73 (1994): 62–78.

Kwong, Julia. *The Political Economy of Corruption in China*. Armonk, NY: M. E. Sharpe, 1997.

Lall, Sanjaya. *Learning from the Asian Tigers: Studies in Technology and Industrial Policy*. Houndmills: Macmillan, 1996.

Larson, James F. *The Telecommunications Revolution in Korea*. Hong Kong: Oxford University Press, 1995.

Lieberthal, Kenneth. *Governing China: From Revolution Through Reform*. New York: Norton, 1995.

—— and Oksenberg, Michel. *Policy Making in China: Leaders, Structures, and Processes*. Princeton, NJ: Princeton University Press, 1988.

Liu, Huaide, and Hu, Hanhui. 'Lun chuangxin shangpin xiaofei de baofaxing' (On the Eruptiveness of Consumption of Creative Goods). *Caijing yanjiu* (The Study of Finance and Research) 7 (2001): 9–16.

Low, Brian. 'The Evolution of China's Telecommunications Equipment Market: A Contextual, Analytical Framework'. *Journal of Business and Industrial Marketing* 20 (2005): 99–108.

Lu, Ding, and Wong, Chee Kong. *China's Telecommunications Market: Entering a Competitive Age*. Cheltenham: Edward Elgar, 2003.

Lü, Xiaobo. *Cadres and Corruption: The Organizational Involution of the Chinese Communist Party*. Stanford, Calif.: Stanford University Press, 2000.

Lynch, Daniel C. *After the Propaganda State: Media, Politics, and 'Thought Work' in Reformed China*. Stanford, Calif.: Stanford University Press, 1999.

—— 'Dilemmas of "Thought Work" in *Fin-de-Siècle* China'. *China Quarterly* 157 (1999): 173–201.

Ma, Hong, and Sun, Shangqing. 'Youdian tongxinye zhongda fangzhen zhengce' (Major Principles and Policies of the Post and Telecommunications Industry). In Hong Ma, and Shangqing Sun, eds. *Xiandai zhongguo jingji da shi dian* (Compendium on Modern China's Economy). Beijing: Zhongguo caizheng jingji chubanshe, 1993: 1852–5.

McIntyre, Bryce T. 'China's Use of the Internet: A Revolution on Hold'. In Paul S. N. Lee, ed. *Telecommunications and Development in China*. Cresskill, NJ: Hamption Press, 1997: 149–69.

McMaster, Susan. *The Telecommunications Industry*. Westport, Conn.: Greenwood Press, 2002.

Masuyama, Seiichi. 'The Evolving Nature of Industrial Policy in East Asia: Liberalization, Upgrading, and Integration'. In Seiichi Masuyama, *et al.*, eds. *Industrial Policies in East Asia*. Tokyo: Nomura Research Institute, 1997: 3–18.

—— *et al.*, eds. *Industrial Policies in East Asia*. Tokyo: Nomura Research Institute, 1997.

Meisner, Maurice. *Mao's China and After*. New York: Free Press, 1999.

MFC Insight. *Carrier Analysis: China Telecom*. Hong Kong: MFC Insight, 2003.

Miozzo, Marcela. 'Transnational Corporations, Industrial Policy and the "War of Incentives": The Case of the Argentine Automobile Industry'. *Development and Change* 31 (2000): 651–80.

Moore, Thomas G. *China in the World Market: Chinese Industry and International Sources of Reform in the post-Mao Era*. Cambridge: Cambridge University Press, 2002.

Mueller, Milton, and Tan, Zixiang. *China in the Information Age*. Westport, Conn.: Praeger Publishers, 1997.

Mundy, Paul, and Sultan, Jacques. *Information Revolutions*. Wageningen, Netherlands: CTA, 2001.

Naughton, Barry. *Growing Out of the Plan: Chinese Economic Reform, 1978–1993*. New York: Cambridge University Press, 1995.

Nie, Norman, and Erbring, Lutz. *Internet and Society: A Preliminary Report*. Stanford, Calif.: Stanford Institute for the Quantitative Study of Society, 17 Feb. 2000.

—— and Hillygus, D. Sunshine. 'The Impact of Internet Use on Sociability: Time-Diary Findings'. *IT&Society* 1 (2002) (pub. online at http://www.stanford.edu/group/siqss/itandsociety/v01i01/v01i01a01.pdf).

—— and Hillygus, D. Sunshine. 'Where Does Internet Time Come From? A Reconnaissance'. *IT&Society* 1 (2002) (pub. online at http://www.stanford.edu/group/siqss/itandsociety/v01i02/v01i02a01.pdf).

Noam, Eli. *Telecommunications in Europe*. New York: Oxford University Press, 1992.

Bibliography

Oi, Jean. 'Fiscal Reform and the Economic Foundations of Local State Corporatism in China'. *World Politics* 45 (1992): 99–126.

Organization for Economic Co-operation and Development (OECD). 'Understanding the Digital Divide'. 2001, pub. online at: http://www.oecd.org/pdf/M00002000/M00002444.pdf.

Oslin, George P. *The Story of Telecommunications*. Macon, Ga.: Mercer University Press, 1992.

Pack, Howard, and Saggi, Kamal. 'Is There a Case for Industrial Policy? A Critical Survey'. *The World Bank Research Observer* 21 (2006): 267–97.

Parker, Edwin B. 'Economic and Social Benefits of the REA Telephone Loan Program'. Mountainview, Calif.: Equatorial Communications, 1981.

—— and Hudson, Heather E. *Electronic Byways*. Boulder, Colo.: Westview Press, 1992.

Pearson, Margaret. *Joint Ventures in the People's Republic of China*. Princeton, NJ: Princeton University Press, 1991.

—— *China's New Business Elite: The Political Consequences of Economic Reform*. Berkeley, Calif.: University of California Press, 1997.

Perry, Elizabeth. *Shanghai on Strike*. Stanford: Stanford University Press, 1993.

Petrazzini, Ben A. *The Political Economy of Telecommunications Reform in Developing Countries*. Westport, Conn.: Praeger, 1995.

Prybyla, Jan. *The Chinese Economy*. Columbia, SC: University of South Carolina Press, 1978.

Pyramid Research. *Telecommunications Markets in China*. Cambridge, Mass.: The Economist Intelligence Unit, 1999.

Qiu, Jack Linchuan. 'Coming to Terms with Informational Stratification in the People's Republic of China'. *Cardozo Arts and Entertainment* 20 (2002): 157–80.

Reischauer, Edwin O., and Craig, Albert M. *Japan: Tradition and Transformation*. Boston, Mass.: Houghton Mifflin, 1989.

'Report of the Working Party on the Accession of China', Addendum, Schedule CLII—The People's Republic of China. Online at http://www.wto.org/English/thewto_e/acc_e/completeacc_e.htm#chn, report label WT/MIN(01)/3/Add.2.

Research Center for Regulation and Competition, Chinese Academy of Social Sciences. 'Universal Service Obligations in China's Telecom Sector: Situations, Reforms, and Implementations'. (May 2002): 1–80. Posted at http://wbln0018.worldbank.org/ppiaf/activity.nsf/files/A070100-S-TCI-RF-CN-FRE.pdf/$FILE/A070100-S-TCI-RF-CN-FRE.pdf.

Richman, Barry. *Industrial Society in Communist China*. New York: Random House, 1969.

Röller, Lars-Hendrick, and Waverman, Leonard. 'Telecommunications Infrastructure and Economic Development: A Simultaneous Approach'. *American Economic Review* 91 (2001): 909–23.

Roseman, Daniel. 'The WTO and Telecommunications Services in China: Three Years On'. *Info* 7 (2005): 25–48.

Bibliography

Sato, Harumasa, and Stevenson, Rodney. 'Telecommunications in Japan'. *Columbia Journal of World Business* 24 (1989): 31–41.

Segal, Adam. *Digital Dragon: High Technology Enterprises in China*. Ithaca, NY: Cornell University Press, 2003.

—— and Thun, Eric. 'Thinking Globally, Acting Locally: Local Governments, Industrial Sectors, and Development in China'. *Politics and Society* 29 (2001): 557–88.

Shanghai Statistical Yearbook. Beijing: China Statistics Press, var. years from 2004.

Shen, Xiaobai. *The Chinese Road to High Technology*. New York: St Martin's Press, 1999.

Solinger, Dorothy. *From Lathes to Looms: China's Industrial Policy in Comparative Perspective, 1979–1982*. Stanford, Calif.: Stanford University Press, 1991.

Sommers, Paul, and Carson, Daniel. 'What the IT Revolution Means for Regional Economic Development'. Washington, DC: Brookings Institution Center on Urban and Metropolitan Policy (Feb. 2003): 1–54.

Statistical Yearbook of China, 1981 (English ed.). Hong Kong: Economic Information & Agency, 1982.

Statistical Yearbook of Shanghai. Beijing: China Statistics Press, various years to 2003.

Steinfeld, Edward S. *Forging Reform in China: The Fate of State-owned Industry*. Cambridge: Cambridge University Press, 1998.

Sun, Li. 'Huawei: Tulang Xiang Shizi de Yanjiang' (The Evolution from a Wolf to a Lion). *IT Jingli Shijie* (CEO and CIO Information Times) 10 (2002): 50–62.

Swaminathan, C. S. 'Role of Telecommunications in National Economy'. In *Telecommunications for National Development*. Calcutta: India Chamber of Commerce, 1983: 67–87.

Tan, Zixiang. 'China's Information Superhighway: What Is It and Who Controls It?' *Telecommunications Policy* 19 (1995): 721–31.

—— 'Internet in China'. *Pacific Telecommunications Conference Proceedings* (Jan. 1996): 621–6.

—— 'The Outlook for the Future of China's CO Switch Market'. *Telecom Asia* 7 (1996): 40–5.

—— 'Regulating China's Internet: Convergence Toward a Coherent Regulatory Regime'. *Telecommunications Policy* 23 (1999): 261–76.

—— 'Product Cycle Theory and Telecommunications Industry—Foreign Direct Investment, Government Policy, and Indigenous Manufacturing in China'. *Telecommunications Policy* 26 (2002): 17–30.

T'ang, Leang-li, ed. *Reconstruction in China*. Shanghai: China United Press, 1935.

Tang, Tinglong. 'The Development of the Shanghai Local Telephone Service'. *China Telecommunications Construction* 1 (1989): 43–7.

Taubman, Geoffry. 'A Not-So World Wide Web: The Internet, China, and the Challenges to Nondemocratic Rule'. *Political Communication* 15 (1998): 255–72.

Tian, Gang. *Shanghai's Role in the Economic Development of China*. Westport, Conn.: Praeger, 1996.

Bibliography

Today's Shanghai P&T. Shanghai: Shanghai Posts and Telecommunications Authority, 1994.

Toulmin, Llewellyn, *et al.* 'Telecommunications Sector: Current Status and Future Paths'. World Bank, Global ICT Department (2006): 1–103.

Triolo, Paul. 'China's United Telecommunications Corporation: New Leader, New Organization, Old Problems'. Report for American Embassy in Beijing, May 1996.

—— and Lovelock, Peter. 'Up, Up, and Away—With Strings Attached'. *China Business Review* 23 (1996): 18–29.

Trumbull, Gunnar. *Silicon and the State: French Innovation Policy in the Internet Age*. Washington, DC: Brookings Institution Press, 2004.

Unicom company brochure, 1995.

Ure, John, ed. *Telecommunications in Asia*. Hong Kong: Hong Kong University Press, 1997.

—— 'Telecommunications in China and the Four Dragons'. In John Ure, ed. *Telecommunications in Asia*. Hong Kong: Hong Kong University Press, 1997: 11–48.

USITO China IT Issue Report 1 (July 1997).

Vogel, Ezra. *The Four Little Dragons: The Spread of Industrialization in East Asia*. Cambridge, Mass.: Harvard University Press, 1991.

Walcott, Susan, and Wheeler, James. 'Atlanta in the Telecommunications Age: The Fiber Optic Information Network'. *Urban Geography* 22 (2001): 316–39.

Walder, Andrew. *Communist Neo-Traditionalism*. Berkeley, Calif.: University of California Press, 1986.

Wang, Shu-hwai. 'China's Modernization in Communications'. In Chi-ming Hou, and Tzong-shian Yu, eds. *Modern Chinese Economic History*. Taipei: Institute of Economics, Academia Sinica, 1977: 335–51.

Wang, Xiangdong. *Xinxihua: Zhongguo 21 shiji de xuanze* (Informatization: China's Choices in the 21st century). Beijing: Shehuikexue wenzhai chubanshe, 1998.

White, Lynn, T. 'All the News: Structure and Politics in Shanghai's Media Reform'. In Chin-Chuan Lee, ed. *The Interplay of Politics and Journalism*. New York: The Guilford Press, 1990: 88–110.

Who's Who in China Current Leaders. Beijing: Foreign Languages Press, 1994.

Wiley, Richard E. 'Competition and Deregulation in Telecommunications: The American Experience'. In Leonard Lewin, ed. *Telecommunications in the U.S.: Trends and Policies*. Dedham, Md.: Artech House, 1981: 37–59.

World Bank. *The East Asian Miracle: Economic Growth and Public Policy*. New York: Oxford University Press, 1993.

Wu, Wei. 'Great Leap or Long March: Some Policy Issues of the Development of the Internet in China'. *Telecommunications Policy* 20 (1996): 699–711.

Wu, Yuan-li. *An Economic Survey of Communist China*. New York: Bookman Associates, 1956.

Xu, Ante, and Armstrong, Phillip. *Chinese Telecom Market*. New York: Northern Business Information, 1995.

Xu, Li. 'Chaju, wenti, duice: dui woguo tongxin qu cheng fazhan bu pingheng de sikao' (Disparities, Problems, and Countermeasures: Thoughts on China's Uneven Regional Communications Development). *Zhongguo ruankexue* (Journal of China Soft Science) (May 1997): 44–51.

Yan, Xu, and Pitt, Douglas. *Chinese Telecommunications Policy*. Boston, Mass.: Artech House, 2002.

Yang, Guobin. 'Environmental NGOs and Institutional Dynamics in China'. *China Quarterly* 181 (2005): 46–66.

Yu, Frederick T. C. 'Communications and Politics in Communist China'. In Lucian Pye, ed. *Communications and Political Development*. Princeton, NJ: Princeton University Press, 1963: 259–97.

Zhang, Bing. 'Understanding China's Telecommunications Policymaking and Reforms'. *Telematics and Informatics* 19 (2002): 331–49.

Zhang, Duanquan. 'Development and Modernization of Telecommunications in New China'. *China Telecommunications Construction* 1 (1989): 4–14.

Zhongguo jiaotong nianjian (China Communications Yearbook). Beijing: Zhongguo jiaotong nianjian she, various years.

Zhongguo jindai youdian shi (Modern History of China's Posts and Telecommunications). Beijing: Renmin youdian chubanshe, 1984.

Zhongguo tongji nianjian (China Statistical Yearbook). Beijing: Zhongguo tongji chubanshe, print and CD-ROM versions, various years.

Zhongguo tongxin nianjian (China Telecommunications Yearbook). Beijing: Zhongguo tongxin nianjian bianjibu, 2004.

Zhongguo tongxin tongji niandu baogao (China Telecommunications Statistics Annual Report). Beijing: Renmin youdian chubanshe, various years.

Zhongguo youdian baike quanshu (Encyclopedia of China's Posts and Telecommunications). Beijing: Renmin youdian chubanshe, 1995.

Zhou, Huasheng, and Kerkhofs, Maurice. 'System 12 Technology Transfer to the People's Republic of China'. *International Journal of Technology Management* 3 (1988): 204–11.

Periodicals, News Agencies, Computer Services, and Websites Cited

AFX Asia
AFX News Limited
Agence France Presse
Alexa Internet Inc. website, http://www.alexa.com
AllAfrica, Inc., Africa News
Asiainfo Daily China News
Asiainfo Daily News
Asia Pacific Telecoms Analyst
Asia Pulse
Associated Press

Bibliography

BBC Summary of World Broadcasts
BBC Worldwide Monitoring
Beijing Review
Bell Canada website, **http://www.bci.ca**
Boston Globe
BT Group web site, **http://www.btplc.com**
Business China, Economist Intelligence Unit
Business Daily Update
Business Times (Singapore)
Businessweek
Businessweek Online, **http://businessweek.com**
Business Wire
China Daily
China Daily website, **http://www.chinadaily.com.cn**
China Economic Review
China Economic Review, Financial Times Information
China Online
China Online website, **http://www.chinaonline.com**
ChinaNex website, **http://www.chinanex.com**
Christian Science Monitor
CNN corporate website, **http://www.cnn.com**
Comtex News Network
Deutsche Presse-Agentur
The Economist
Economist Intelligence Unit, *Business China*
Electronic Buyers News
Far Eastern Economic Review
Financial Times
Financial Times (London)
Financial Times Asia Intelligence Wire
Financial Times Information
Financial Times Information, Asia Africa Intelligence Wire
Financial Times Information, Business Daily Update
Financial Times Information, China Online
Financial Times Information, Global News Wire
Financial Times Information, *Indian Express*
Financial Times Information, International Market Insight Reports
Financial Times Information, Newsbase Russian Daily Bulletin
Financial Times Information, Sinocast China
Financial Times Information, SinoCast China Business Daily News
Financial Times Information, SinoCast China IT Watch
Financial Times Information, SinoCast News Wire
Financial Times, Telenews Asia Newsletter

Bibliography

Forbes website, **http://www.forbes.com**
FT Asia Intelligence Wire, Asian Review of Business and Technology
Guardian (London)
Hong Kong Standard
Hoover's, Inc. website, **http://www.hoovers.com**
Human Rights Watch website, **http://www.hrw.org**
IAMASIA website, **http://www.iamasia.com**
India's Ministry of Commerce and Industry website: **http://www.dipp.nic.in**
Interfax News Agency, China Business
International Herald Tribune
International Telecommunications Union (ITU) website, **http://www.itu.int**
IPS Interpress Service
Jingji Cankao Bao (Economic Information News)
Kyodo News Service
Latin America News Digest
LeadDiscovery website, **http://www.leaddiscovery.com.cn**
Los Angeles Times
Ministry of Information Industry (MII) website, **http://www.mii.gov.cn**
New York Times
People's Daily
Renmin youdian (People's Posts and Telecommunications)
Reuters World Report
Shenzhen Data Communications Bureau website, **http://www.shenzhenwindow.net/topten/e-zte.htm**
Sinocast China IT Report
SinoCast China IT Watch
South China Morning Post
South China Morning Post website, **http://china.scmp.com**
Straits Times (Singapore)
Sydney Morning Herald
Telecommunication Development Asia–Pacific
Telecommunications Development Asia–Pacific website, **http://www.tdap.co.uk**
TIW (Telesystem International Wireless Inc.) website, **http://www.tiw.ca/**
UNESCO's International Bureau of Education website, **http://www.ibe.unesco.org**
United Press International
Universal Service Administrative Company website, **http://www.usac.org**
Wall Street Journal
Wall Street Journal website, **http://online.wsj.com**
Washington Post
Xinhua Economic News Service
Xinhua Financial Network News
Xinhua Financial News
Xinhua Financial News Network

Bibliography

Xinhua General News Service
Xinhua General Overseas News Service
Xinhua News Agency
Xinhua News Service
ZTE corporate website, **http://www.zte.com.cn**

Index

administrative guidance 4
Advanced Research Projects Agency 80
Africa:
  industrial growth in 7
  industrial policy in 184
  telecommunications development in 22, 45, 155, 160, 184
Africa Online 160
Alcatel:
  creation of Alcatel Shanghai Bell by 131–2, 133
  equipment sales of to China 117
  joint venture of in China 118, 120, 131, 141, 142, 155, 156
  research and development spending by 133
  see also Shanghai Bell
Alibaba 103, 104
América Móvil 22
America Online 81
American Telephone and Telegraph Company:
  American Internet development by 80, 81
  breakup of 42, 44, 46, 47
  competition against 19–21, 57, 201–2n110
  equipment manufacture of 114
  growth of 19–21, 25, 160
  investment of in China 118, 150, 155
  investment of in Shanghai 73, 150–2
  joint ventures of in Shanghai 150–2
  problems of in Shanghai 151–2
  revenue of in Shanghai 152
Ameritech:
  as China Unicom partner 54, 56–7, 200n46
Anhui Province:
  Internet penetration in 170–1
  Internet police in 96
  telecommunications development in 168
Argentina:
  automobile industry in 7
  telecommunications industry in 22
ARPA, *see* Advanced Research Projects Agency

ARPANET 80
AsiaInfo 85
Asian American Telecommunications Corporation 56 table 3.4, 64
Asian Development Bank 117
Asian economic crisis of 1997 7–8, 200n69
AT&T, *see* American Telephone and Telegraph Company
automobile industry:
  in Argentina 7
  in China 11, 138

Baark, Eric 26
Baidu.com 102 table 4.4, 103, 105
Bangalore:
  development in 186
Beijing:
  China Unicom in 50, 148
  Internet cafés in 154
  Internet development in 92, 93, 102, 107, 158, 169
  Internet penetration in 170–1
  Internet pornography in 99
  local government power of 138
  mobile phones in 51, 166 table 7.2, 167
  municipal corporations in 145
  per capita income levels in 170, 172
  purchase of school computers in 178–9
  telecommunications equipment manufacture in 113, 114
  telegraph lines in 27
  telephones in 29, 158
Beijing Catch Communications Group 48 table 3.1, 55 table 3.1
Beijing Comprehensive Investment Corporation 120
Beijing Enterprises Holding Ltd. 132
Beijing International Switching Company:
  controlling acquisition by Siemens of 132
  creation of 119–20, 123
  production of 120, 121, 131
  reorganization of 132

231

Index

Beijing Jeep Corporation 11
Beijing Olympics:
   mobile phone standard for 75
Beijing Telecom:
   Internet service of 89
Beijing Telecommunications Authority 120
Beijing Wire Communications Plant 114, 120, 123, 132, 209n47
Beiping, *see* Beijing
Bell Canada:
   as China Unicom partner 54, 56 table 3.4, 64
Bell Telephone Company 19
   see also American Telephone and Telegraph Company
Bell Telephone Laboratories 20
Beto corporation 128
BISC, *see* Beijing International Switching Company
Blogs, *see* Internet in China: blogs in
Boas, Taylor 14
Brazil:
   industrial policy in 9
British Telecom 21, 44
Broadband, *see* Internet in China: broadband services of
BTM 118
bureaucratic rivalry in China:
   within China Unicom 61
   among ministries 47, 78, 93–4, 120
   see also competition in China
Business Week 8

cable television, *see* Internet in China: cable television used for
Capital Online 92, 102
Carlson, Daniel 137
CCF, *see* joint ventures: by China Unicom, and joint ventures: by Shanghai Unicom
CCT Holdings:
   investment of in China Unicom 56 table 3.4, 64, 148, 149, 200n46
CDMA:
   as China Unicom's standard 67, 75
   costs of 67
CDMA-2000 75
Censorship, *see* Internet in China: censorship of
central-local government relations 185–6
   see also central-local government relations in China
central-local government relations in China 11, 29, 138, 152, 185–6

   in Shanghai's telecommunications 135, 136, 139, 141 fig. 6.2, 142, 147, 149, 155, 157
   in telecommunications 31–2, 36, 37
   see also local government
CERNET, *see* China Education and Research Network
CGWNet 86, 87, 88 table 4.1
Chan, Joseph 105
Chang An company 131
Chang, Xiaobing 77
Changcheng Industrial Company 125
Chase, Michael 14
Cheng, Dongsheng 128
Cheng, Tun-jen 5
Cheng, Xiyuan 151
Chiang, Kai-shek 28
China:
   digital divide in 94, 107, 158–9, 161–82
   GDP growth in 37, 38 table 2.1
   industrial policy in, *see* industrial policy in China
   Internet development in, *see* Internet in China
   mobile phones in, *see* mobile phones in China
   Shanghai's telecommunications development 136–7, 139–57
   telecommunications equipment in, *see* telecommunications equipment manufacture in China
   telecommunications history of: 1860s-1911 26–8; 1911–49 28–9; 1949–76 30–4, 113–15, 162–4; 1976–93 34–7, 115–21, 164–5; 1994–2007 38–40, 41, 46–78, 129–34, 165–7
   telecommunications studies of 12–14
   see also central-local government relations in China; digital divide in China; foreign investment in China; industrial policy in China; Internet in China; joint ventures; local government; mobile phones in China; rural telecommunications in China; Shanghai; telecommunications in China; telecommunications equipment manufacture in China; urban telecommunications; *and under specific Chinese government ministries and telecommunications companies*
China Academic Network 83
ChinaCom 48 table 3.1, 54, 55 table 3.3, 56, 61
China.com 102 table 4.4, 104

232

Index

China Education and Research Network 83, 87, 88 table 4.1
  connection of schools and universities by 178, 181
  lack of funding for 178
China Everbright ITIC 48 table 3.1
ChinaGBN, see Jitong: ChinaGBN Internet network of
China Great Wall Computer Group 178
China Huaneng Group 48 table 3.1, 55 table 3.1
China International Trust and Investment Corporation 48 table 3.1, 54, 55 table 3.3, 56 table 3.4
China Merchants Holdings Ltd. 48 table 3.1, 55 table 3.1
China Mobile:
  China Railcom and 70
  competition of 68, 69
  creation of 66, 68
  expansion of to rural areas 180
  Internet network of 87, 88 table 4.1
  leaders of 68, 74, 77
  market share of 65 table 3.5
  overseas investments of 74, 189
  prohibition of joint ventures with 142
  subscribers of 65 table 3.5
  telecommunications standards of 68
  third generation mobile phones and license for 67, 73, 75
  see also China Telecom
China National Chemicals Import and Export Corporation:
  as China Unicom shareholder 48 table 3.1, 61
ChinaNET 85, 87, 88–9, 90, 92, 93, 169
ChinaNET/163 85, 91, 92
China Netcom (old):
  absorption of 72, 87
  creation of 86
  foreign investment in 86, 104
  Internet network of 86, 87, 88 table 4.1
  prepaid cards of 66
China Netcom (new):
  competition of 72
  creation of 72, 86, 87, 180
  Internet network of 87, 88 table 4.1
  lack of a mobile network license 73
  xiaolingtong service of 69
China News Digest 98
China169 87, 88 table 4.1
China Online 91
China Orient Telecom Satellite 71
CHINAPAC 85

China Posts and Telecommunications Industry Corporation:
  AT&T joint venture of 150
  creation of 115–16
  foreign technology absorption of 117
  investment of in Great Dragon 123, 124, 209n47
  Shanghai Bell joint venture of 118, 121, 124, 131, 142
  policies of 116
  reorganization of 131, 133
  see also Putian
China Railcom:
  competition of 69, 70–1, 77
  creation of 70
  problems of 70–1
  see also China Tietong
China Research Network 83, 86
China Resources Group 48 table 3.1, 55 table 3.1
China Satellite Communications:
  creation of 71
  competition of 69, 71, 77
  paging services of 67
China Science and Technology Network 83, 87, 88 table 4.1
China Telecom:
  competition of 16, 53, 58, 59, 63, 66, 69–70, 72, 148
  creation of 50, 143
  GSM standard of 67
  initial public offering of 63, 65
  Internet service of 85, 92
  lack of a mobile network license 69, 73
  leaders of 68, 77
  mobile phone market share of 52 table 3.2, 65 table 3.5
  mobile phone subscribers of 52 table 3.2, 65 table 3.5
  organization of 50
  prepaid cards of 66
  prohibition of foreign investment in 55, 142
  reorganization of in 2000 66, 67
  reorganization of in 2001 72, 87, 177, 180
  universal service fund and 177
  xiaolingtong service of 68–9
  see also China Mobile, ChinaNET
China Tietong:
  competition of 71
  market share of 71
  see also China Railcom
China Unicom:
  CDMA service of 67, 75

233

Index

China Unicom: (*cont.*)
  competition of 16, 43, 50–3, 58, 59, 62, 63, 68, 77
  creation of 37, 48–9, 198n11
  early goals of 49, 62
  fixed-line network of 53, 58, 59, 61, 63
  foreign investment and joint ventures in 50, 53–8, 59, 60, 61, 62, 63–4, 77, 202n131
  growth of 52
  income of 53
  initial public offering of 61, 63, 64, 65, 149, 198–9n15
  Internet network of 86, 87, 88 table 4.1
  leaders of 60–1, 62, 63, 77
  list of joint ventures with 56 table 3.4
  mandate of 49
  market share of 52 table 3.2, 65 table 3.5, 66
  mobile phone equipment for 130–1
  operating problems of 57–8, 59–62
  paging services of 67, 71
  prepaid cards of 66
  regional shareholders of 55 table 3.3
  reorganization of 62–3, 66, 86
  revenue of 52
  Shanghai branch of 147–9
  shareholders of 48 table 3.1
  subscribers of 52, 65 table 3.5, 67
  third generation mobile phones and license for 73, 75
  see also Shanghai Unicom; Unicom Investor Group
China United Telecommunications Corporation, *see* China Unicom
Chinese Academy of Sciences 47, 86
Chinese Communist party:
  central committee of 31, 36, 61
  Fifteenth Congress of 61
  private communications network for 177
  role of in the economy 35
Chinese military, *see* People's Liberation Army
CIETNet 86, 87, 88 table 4.1
Cisco Systems 85
  conflict with Huawei by 132
  sales of in China 132–3
CITIC, *see* China International Trust and Investment Corporation
CITIC Bank 131
'City Smart' 69
civil society:
  in China's Internet community 109, 110–11, 152
Clark, Duncan 126

CMNet 86, 87, 88 table 4.1
competition:
  among American phone companies 19–21, 44
  in China, *see* competition in China
  in developing countries' telecommunications 22, 23, 24–5
  among European phone companies 44
  among foreign investors in China 57, 60, 73
  industrial policy and 5, 16, 43, 184, 188, 189
  among Japanese phone companies 44–5
  in London's telecommunications market 137
  among Mexican phone companies 22
  among South Korean phone companies 24–5
  in telecommunications development model 25
  see also competition in China
competition in China:
  among Internet content providers 79, 103–5
  among Internet networks 85–8
  among Internet service providers 90–2, 93–4, 95, 180
  among ministries 47, 58, 61, 76, 78, 120
  at the municipal level 138
  among phone companies 42, 43, 45–7, 49–53, 58, 59, 62, 66, 68, 69–71, 73, 75–7, 180
  in Shanghai's telecommunications industry 136, 144, 147, 148–9, 152, 153–4, 155, 156
  among telecommunications equipment companies 113, 119, 120, 121, 123, 124, 125, 128, 131, 132–4
  overall effect of on telecommunications 184, 188, 189
'*cun cun tong dianhua*' policy 165–9
Confucianism:
  in industrial policy 10
Construction Bank of China 131
Corning International 150
corruption:
  in China 147, 186–7
  in India 187
  relation of to industrial policy 186
  in Shanghai's telecommunications development 147, 156–7, 214n68
Cortell, Andrew 9
CRNET 71, 86, 87, 88 table 4.1
CSNet 86–7, 88 table 4.1
CSTNet, *see* China Science and Technology Network

Index

Cultural Revolution:
  development of telecommunications in 32–3, 115, 139, 163–4

Daewoo 56 table 3.4
Dahrendorf, Ralf 4
danwei 33
Datang Telecom Technology 75
Dell 86
Deng, Xiaoping 18, 35
  policies of 32, 139, 164
depreciation:
  in China's telecommunications sector 38
Deutsche Telecom 56 table 3.4
digital divide:
  definition of 161
  in Africa 160
  in China, *see* digital divide in China
  in India 160
  policies to address 160–1
  studies of 159–61
  in the US 160
  see also digital divide in China
digital divide in China:
  definition of 161–2, 215n14
  for Internet services 94, 107, 111, 158, 159, 168, 169–73, 174–6, 178–9, 180–2
  new technologies to alleviate 167, 179–81
  policies to address 159, 161, 162–70, 176–9, 180, 181, 187
  problems of solving 166–7, 174–6, 178
  relation of to education levels 174–5, 217n59
  for telephone services 158, 159, 162–9, 176–7, 179–80, 181–2
  universal service fund for 176, 177, 180
  see also digital divide; rural telecommunications in China; urban telecommunications
Ding Guan'gen 96
'Directive 56', *see* State Council: 'Directive 56'
'Directive 178', *see* State Council: 'Directive 178'
Directorate General of Telecommunications (China) 50
dissidents:
  use of China's Internet by 14, 95, 96–8, 99, 100, 107, 109, 110–11, 175–6
  use of China's telecommunications network by 46, 176
DoCoMo, *see* NTT DoCoMo
Dore, Ronald 4

Eachnet 103, 104
Eastday.com 102, 137, 145, 152–5, 156
  creation of 153
  Internet cafés of 153–4
  municipal assistance for 153–4
  web site of 153
eBay 103, 104
electronic commerce in China: 102–5
  'B2B' commerce of 103–4
  'C2C' commerce of 103–4
  foreign companies in 103, 104–5
  see also under specific electronic commerce companies
embedded autonomy 9
Erbring, Lutz 107–8
Ericsson;
  equipment sales of to China 117
Evans, Peter:
  theories of 6–7, 9, 10, 186

Falun Gong 100
Federal Communications Commission 20, 26, 47
feng shui 26
First Ministry of Machine Building 113–14
forced labour:
  in China's telecommunications network building 31
foreign investment:
  in China, *see* foreign investment in China
  in industrial policy 6, 7, 184, 185, 188–9
  in Japan 5, 185
  role of in developing countries' telecommunications 21, 22, 23, 25, 156, 157
  role of in Japan's telecommunications 24
  in South Korea 185
  see also foreign investment in China
foreign investment in China 188–9
  in China's Internet 86, 100, 103–5, 188
  in China Unicom 53–8, 59, 60, 61, 62, 63–4, 77
  control of telegraph lines by 29
  exception of rules for AT&T 151–2
  outward investment by China Mobile 74
  outward investment by Huawei 128
  role of in China's telecommunications services sector 73, 78, 155, 168, 184, 185, 188–9; government rules on 2, 41–2, 46, 49–50, 54, 58–9. 63–4, 77
  in Shanghai's telecommunications industry 136, 141–2, 143–4, 147–52, 155, 156, 157

235

Index

foreign investment in China (cont.)
  in telecommunications equipment
    manufacturing 112, 116, 117–22, 123,
    130–2, 133–4, 164, 188, 198n12
  under World Trade Organization
    rules 73–4, 188
  see also American Telephone and Telegraph
    Company; Beijing International Switching
    Company; China Unicom; joint ventures;
    Shanghai Bell; Shanghai Unicom; Tianjin
    NEC; World Trade Organization; and under
    other specific companies
France:
  Internet in 44, 81, 90 table 4.2
  investment of in China 37
  loans of to China 121
  municipal telecommunications in 137
  telecommunications history of 21, 44
  see also Alcatel; France Telecom; Shanghai
    Bell
France Telecom 21, 22, 44, 137
  as China Unicom partner 56 table 3.4, 64
  Internet service of 81
Fujian Province:
  mobile phones in 166 table 7.2, 167
Fujitsu:
  equipment sales of to China 117, 127, 143
Fuzhou:
  telegraph lines in 26

Gansu Province:
  Internet penetration in 170–1
  mobile phones in 166 table 7.2, 167
GDP, see Gross Domestic Product
general packet radio service 68
gender gap, see Internet in China: gender gap of
General Telecommunications
  Administration 33
Germany:
  investment of in China 37
  see also Siemens; Beijing International
    Switching Company
Ghana:
  telecommunications development in 160
globalization:
  challenge of to China 2
Global System for Mobile:
  as telecommunications standard in
    China 67, 68
Godement, François 8
Goldman Sachs 86
Google:
  Internet service of in China 100, 103,
    104, 105

Graham, Stephen 137
Great Britain:
  municipal telecommunications in 137
  telecommunications history of 21, 44
Great Dragon Group:
  creation of 122–3, 124, 209n47
  loans for 123
  partners in 123, 209n47
  problems of 124
  production of 123, 124, 125
Great Firewall of China, see Internet in
  China: blocking of web sites in
Great Leap Forward:
  development of telecommunications
    in 32, 162–3, 167
Great Northern Telegraph Company 26,
  27, 139
Gross Domestic Product:
  growth of in China 37, 38 table 2.1,
    212n18
  per capita in Chinese cities 92
GSM, see Global System for Mobile
GTE Sprint 201–2n110
Guangdong Province 107
  China Unicom in 53
  Internet cafés in 96
  Internet penetration in 170 table 7.3
  Internet service in 90, 169
  mobile phones in 166 table 7.2, 167
  telecommunications development in 168
  telephones (fixed-line) in 29
Guangxi:
  Internet penetration in 172, 175
Guangzhou:
  China Unicom in 50, 60, 67, 148
  Internet cafés in 154
  local government power of 138
  mobile phones in 51
  municipal corporations in 145
Guangzhou Peugeot 11
Guizhou Province:
  illiteracy in 174
  Internet development in 107, 158
  Internet penetration in 170–1
  mobile phones in 166 table 7.2, 167
  per capita income levels in 172
  telegraph lines to 27
Guoxin Paging 67

Hackler, Darrene 137
Haggard, Stephan 5
Hankou:
  telephones in 29
Hardy, Andrew 214n2

236

Index

Henan Province:
  Internet penetration in 170 table 7.3
  mobile phones in 166 table 7.2
Hillygus, D. 109
Hong Kong:
  industrial policy in 5, 6, 9
  telegraph lines to 26
Hu, Angang 168–9
Hu, Hanhui 168
Hu, Qili 47, 48, 58, 61, 63, 198n12
Hua, Guofeng 34, 35
Hua Xin company 131
Huang, Qi 97
Huatong, *see* ChinaCom
Huawei Technologies 123, 124, 125, 126, 183
  conflict with Cisco Systems by 132–3
  creation of 126–7
  expansion of 127–8, 133, 156
  exports by 128, 133, 187, 189
  Internet equipment production of 132–3
  market share of 128, 132, 133
  overseas investment of 127, 128
  private status of 209n65
  production of 127, 128, 132–3, 134
  regional shell companies of 128–9, 187
  research and development spending by 127, 128, 133
  revenue of 128, 133
  strategies of 127–9, 132–3
  technology transfer to 127
Hudson, Heather:
  theories of on rural telecommunications 159–61, 167, 214n2

ICP, *see* Internet in China: content provision for
illiteracy:
  as a problem for Internet access 162, 174
  special phones adapted to 180
Imperial Telegraph Administration 27–8
income levels in China:
  disparities in 172 table 7.4, 174
India:
  case of Bangalore 186
  case of Kerala 186
  corruption in 187
  democracy in 188
  foreign investment in 155, 185, 186, 194n31
  industrial growth in 7
  industrial policy in 9, 184, 185, 186
  Internet development in 82, 90 table 4.2
  Internet usage costs in 94
  telecommunications: costs in 76 table 3.6; development in 45, 160, 184, 186; history of 22–3
  telephones and teledensity in 39 table 2.2
industrial policy:
  assessment of 8–9, 10
  at the municipal level in telecommunications 135, 137, 156, 157
  central-local government relations and 185–6
  competition and 5, 16, 43, 184, 188, 189
  corruption and 186–7
  definition of 4, 190n2
  democracy and 187–8
  foreign investment and 6, 7, 184, 185, 188–9
  in China, *see* industrial policy in China
  in India 9, 184, 185, 186
  in Japan 4–5, 9, 15, 23, 25, 120, 185
  in South Korea 5, 6, 7, 9, 25, 185
  industry-specific focus on 14–15
  pitfalls of 5, 7–9
  priorities of 187–8
  public benefits and 183
  role of in development 3–12, 14–17
  seminal theoretical works of 3–10, 186
  theoretical corollaries for from the case of China's telecommunications 183–9
  see also industrial policy in China; *and under other specific countries*
industrial policy in China 10–12, 15–17
  of China's telecommunications ministries, *see* Ministry of Posts and Telecommunications: policies of, *and* Ministry of Information Industry: policies of
  corollaries to industrial policy theory from 183–9
  democracy and 187–8
  failure of in Great Dragon case 124
  in Shanghai's telecommunications industry 136, 142–6, 147, 155, 156, 157, 185–6; in Eastday.com case 152, 153–4; in foreign-invested projects 148, 149, 151
  of the State Council, *see* State Council: policies of
  in telecommunications equipment manufacture 121–2, 123, 124, 126, 134
  in Mao-era telecommunications development 30–2, 33, 34,
  in post-Mao telecommunications development 35, 36–7, 38–40, 41–2
  see also central-local government relations; competition in China; foreign investment in China; industrial policy
InfoHighway 90, 91, 92, 93, 97
information control in China 58, 164
  in the Internet 2, 14, 79, 95–100, 105, 111, 175–6

237

Index

information control in China (*cont.*)
   in the Qing dynasty 26
   in Shanghai 152, 154
   see also Internet in China: dissident use of, *and* Internet in China: self-censorship in
initial public offering:
   of Baidu 103
   of China Telecom 63, 65
   of China Unicom 61, 63, 64, 65, 149, 198–9n15
   lack of for China Tietong 71
   lack of for Huawei 129
   of ZTE 125
Inner Mongolia:
   mobile phones in 166 table 7.2
Institute of High Energy Physics 83
intellectual property rights:
   conflict over in Internet equipment 132
   protection of in China's telecommunications equipment sector 121, 126, 131–2, 156, 188
interconnection:
   in China's telecommunications network 52, 53, 57, 63, 70
Internet:
   development of: in China, *see* Internet in China; in France 44, 81; in India 82; in Japan 81–2; in the US 80–1
   social isolation of 108
   telephones and teledensity in 39 table 2.2
   usage costs in selected Asian nations 94 table 4.3
   usage patterns of in the US 107–9
   user numbers and penetration in selected nations 90 table 4.2
   see also Internet in China
Internet cafés, *see* Internet in China: cafés of; Shanghai: Internet cafés in
Internet in China:
   addiction to 96, 109
   anti-Japan student protests and 110
   bandwidth of 79, 87, 88 table 4.1
   blocking of web sites in 98–9, 100, 205n48
   blogs in 98, 108 table 4.5
   broadband services of 72–3, 87, 93, 153, 158, 180, 183, 205n38
   cable television used for 94, 153, 180
   cafés of 96–7, 109, 153–4
   censorship of 14, 95–6, 97, 98–100, 152, 154
   chat groups of 97, 108 table 4.5, 109–10, 206n91
   civil society and 109, 110–11, 152
   commercial activity in 100–5
   construction of network for 82–8
   content provision for 79, 94–104, 111, 153
   control of by government 2, 79, 85, 94–5, 95–100, 105, 111
   costs of usage for 92–3, 94, 169–70, 172, 178, 180
   demographics of 105
   digital divide in, *see* digital divide in China: for Internet services
   dissident use of 14, 95, 96–8, 99, 100, 107, 109, 110–11, 175–6
   domain names of 79, 101, 152
   educational service of 159, 178–9
   effect of World Trade Organization rules on 73
   electronic commerce in 102–5, 108 table 4.5, 111
   foreign influence on 40–1, 80, 98–9, 100
   foreign investment in 86, 100, 103–5
   gambling in 95, 98, 109
   games in 103, 108 table 4.5, 109, 154
   gender gap of 106–7
   influence on government policies of 110, 206n91
   major networks of 84 fig. 4.1
   modem dial-up for 72, 85, 91, 93, 153, 170, 174, 205n38
   mobile services of 67, 68, 75, 103
   network structure of 79, 83–94, 169
   penetration of 90 table 4.2, 170–3
   policing of 96, 97, 109, 175–6
   popular organizing using 99, 100, 109, 110–11, 175–6, 206n68
   popular web sites of 101–2, 103, 104, 105, 153
   pornography in 95, 96, 98, 99, 109
   power lines used for 180–1
   private Chinese companies in 90, 92, 93, 101–5, 111
   regional penetration of 170–6, 178–9, 180–2
   regulations on usage of 95, 96, 97, 98, 99, 105, 111
   self-censorship in 97, 104, 110, 111, 175–6
   service providers of 86, 88–94, 111, 153
   service provision hierarchy of 91 fig. 4.3
   Shanghai's development in 86, 93, 102, 107, 136–7, 140 fig. 6.1, 144–5, 146, 151, 152–5
   slowing growth of 171
   social isolation of 108
   studies of 13–14
   studies of by Chinese scholars 14
   telecommunications equipment for 129, 130, 132–3

Index

usage by age 106, 107
usage patterns of 107–11
user, definition of 89 fig. 4.2, 204n25
user demographics of 79
users, number of 1, 79, 89, 130 fig. 5.3, 158, 204n25; in Shanghai 136
World Trade Organization requirements of 104, 105
youth and 96, 106, 109
see also under specific Internet companies
Interstate Commerce Commission 19–20
IPTV, *see* Internet in China: television used for
ISP, *see* Internet in China: service providers of
Itochu 56 table 3.4, 64
ITT Corporation 118

Japan:
dual economy of 15
economic ties with China in 1960s 32
foreign investment in 5, 185
industrial policy in 4–5, 9, 15, 23, 25, 120, 185
Internet development in 81–2
investment of in China 37, 56 table 3.4, 64, 118, 120, 121
mobile Internet in 82
mobile phones in 45, 82
occupation of China by 29, 34, 113
relations of with China 110
student protests against in China 110
telecommunications history of 23–4, 44–5, 208n37
see also Fujitsu; Nippon Electric Company; Nippon Telegraph and Telephone Corporation; Tianjin NEC
Jiang, Mianheng:
in China's Internet development 86
in China's telecommunications development 186–7
in SAIL 145
in Shanghai's telecommunications development 86, 145, 153,156–7, 213–14n67
Jiang, Zemin 86, 143, 145, 156, 187, 213–14n67
and China's telecommunications policy 127–8, 161
Jiangsu Netcom 72
Jiangsu Province:
mobile phones in 166 table 7.2
Jitong:
ChinaGBN Internet network of 85–6, 88 table 4.1, 90, 92
prepaid cards of 66
reorganization of 86, 87
Johnson, Chalmers:
theories of 4–5, 9, 10, 16, 183, 189
joint ventures:
in China's telecommunications equipment manufacturing 117–20, 121–2, 125, 126, 129–32, 133, 134, 150
in China under World Trade Organization rules 73
by China Unicom 54–8, 59, 60, 61, 62, 77, 148–9, 202n131
by China Unicom, cancellation of 63–5, 149
by Huawei 128–9
policies for in Shanghai 143–4
by Shanghai Unicom 148–9
see also American Telephone and Telegraph Company; Beijing International Switching Company; China Unicom; foreign investment in China; Shanghai Bell; Shanghai Unicom; Tianjin NEC; *and under other specific companies*
Julong, *see* Great Dragon Group

Kalathil, Shanthi 14
Katz, Richard 15
KDDI 45
Kelly, George 4
Kenya:
telecommunications industry in 22, 45, 160
Kerala:
development in 186
Kingsbury Commitment 20
Kissinger, Henry 60
KMT, *see* Republican-era government
Kokusai Denshin Denwa 24
Korea, South:
foreign investment in 185
growth in 8
industrial policy in 5, 6, 7, 9, 25, 185
Internet users in 90 table 4.2
telephones and teledensity in 39 table 2.2
telecommunications history of 24–5
Korea Telecom 24
Korea Telecommunications Authority 24
Krugman, Paul 7
Kuomintang, *see* Republican-era government

Lall, Sanjaya 6, 7
Latin America:
industrial policy in 6–7
Leading Group for the Revitalization of the Electronics Industry 36

239

Index

Li, Huifen 60-1, 63
Li, Peng 36, 40, 46, 48
Liantong, see China Unicom
Liaoning Province:
  Internet service in 90, 169
  mobile phones in 166 table 7.2
licences:
  for telecommunications companies in China 60, 63, 69, 70, 71, 73, 75, 77
Lieberthal, Kenneth 11
Lin, Hai 97
'Little Smart', see *xiaolingtong*
Liu, Dingzhuan 123
Liu, Huaide 168
Liu, Jianfeng 60, 62, 200n70
Liu, Lili 128
Liu, Shaoqi 32
Liu, Zhenyuan 146, 148, 149
loans:
  for America's telecommunications sector 20, 160
  for China's telecommunications sector 36, 38, 53, 117, 123, 164
  for Huawei 127
  for Japan's telecommunications sector 23
  for Shanghai Bell 121, 131
local government:
  in China 12, 137-8
  in China's telecommunications development 31-2, 36, 37, 50, 102, 163
  in China's Republican era 29
  control of media by in Shanghai 152, 153, 154
  in France's telecommunications development 137
  local state corporatism of in China 138, 155
  in London's telecommunications development 137
  role of in Shanghai's telecommunications development 136, 142-6, 148, 151, 155, 156, 157, 185-6; in Eastday.com 152, 153-4
  in the US' telecommunications development 137
  see also central-local government relations in China; provincial telecommunications administrations; Shanghai; *and under specific cities and provinces*
long-distance communication in China:
  China Unicom and 49, 52, 58, 149
  history of 29, 30, 33, 139
Lucent Technologies 44, 150
Lu, Ding 13

Lü, Xiaobo 147
Lu, Xinkui 62
Luo, Gan 200n69
Lynch, Daniel 11, 13-14, 198n12

McCaw International:
  as China Unicom partner 56 table 3.4, 57, 148, 149, 155, 200n46
McIntyre, Bryce 13
Malaysia:
  growth in 8
  industrial policy in 6, 9
  Internet usage costs in 94
  telecommunications costs in 76 table 3.6
Manchuria 28
Mao, Zedong 18, 33, 34, 46, 139, 164
  on telecommunications 30
mass communication:
  in pre-reform China 30
MasterCall 56 table 3.4
Masuyama, Seiichi 7-8
MCI 81
MEI, *see* Ministry of Electronics Industry
Meisner, Maurice 215n17
Mercury Communications 44
Metromedia 64
Mexico:
  telecommunications history of 21-2
Microsoft:
  Internet service of in China 100, 102 table 4.4, 104
MII, *see* Ministry of Information Industry
Millicom International Cellular 74
Minitel 81
Ministry of Aerospace Industry 124, 125
Ministry of Communications (China) 29, 30, 195n62
Ministry of Communications (Japan) 23
Ministry of Communications (South Korea) 24
Ministry of Education 178, 181
Ministry of Electric Power:
  as China Unicom shareholder 48, 49, 53, 55 table 3.3, 58, 59, 60, 61, 69-70
  in China's telecommunications development 47
Ministry of Electronics Industry:
  as China Unicom shareholder 48, 50, 55 table 3.3, 60, 61, 63, 198n12
  in China's telecommunications development 47
  Internet network of 85-6
  investment of in BISC 120, 121, 124
  investment of in Great Dragon 123, 124, 209n47

Index

reorganization of 62, 86
role of in telecommunications equipment manufacture 123
Ministry of Finance 177
Ministry of Information Industry:
  creation of 62, 86, 131, 149
  financial stake of in Shanghai Bell 131
  Internet policies of 86, 87, 88, 92, 93, 100–1, 104, 111
  leaders of 40, 62, 63, 77
  policies of 63, 64, 69, 70, 74, 77, 82, 144, 146; on telecommunications equipment manufacture 126, 129, 133; on rural telecommunications 177, 181
  see also Ministry of Posts and Telecommunications (China)
Ministry of International Trade and Industry:
  in Japan's development 4–5
Ministry of Posts and Communications (China) 28
Ministry of Posts and Telecommunications (China):
  competition of with China Unicom 48, 49, 148–9, 198n12
  control of Internet network by 82–3, 85
  cooperation of with AT&T 150–1
  funding for 38–9
  history of 1949–78 30–3
  Internet policies of 90, 92
  investment of: in BISC 120, 121, 132; in Great Dragon 123, 124, 209n47; in Shanghai Bell 118, 121, 124, 131; in Tianjin NEC 120
  leaders of 32, 35–6, 40–1
  policies of 36, 37, 43, 46, 56, 82, 142; against China Unicom 49–50, 52, 53, 54, 57, 58; for rural telecommunications 162, 164, 165–6, 167–8; in Shanghai 142, 150
  profits of 32
  regulatory powers of 31, 47, 59
  reorganization of 62, 131, 149
  role of in telecommunications equipment manufacture 113, 114–16, 118, 123, 125
  separation of regulation and operation in 50
  see also Ministry of Information Industry
Ministry of Posts and Telecommunications (Japan) 81–2
Ministry of Power, *see* Ministry of Electric Power
Ministry of Public Security:
  regulations of for China's Internet use 96, 205n48

Ministry of Radio, Film, and Television 62
Ministry of Railways:
  and China Railcom 70–1
  in China's telecommunications development 47
  and China Tietong 71
  as China Unicom shareholder 48, 49, 53, 55 table 3.3, 58, 59, 60, 61, 69–70
  fixed-line network license for 70
  Internet network of 86, 87
  investment of in China Netcom (old) 86
  merger talks of with China Unicom 70
Miozzo, Marcela 7
MITI, *see* Ministry of International Trade and Industry
mobile phones in China:
  costs of 51, 66, 67, 69, 76 table 3.6, 179, 180
  costs of in Shanghai 144
  government control of 110, 176
  growth of in Shanghai 51, 136, 140 fig. 6.1, 144, 148–9
  handset sales of 133–4, 180
  Internet use of 67, 68, 75, 103
  licenses for networks of 60, 63, 69, 73, 75
  network growth of 52, 65–6, 67, 68, 69, 71, 73, 75, 76, 77
  number of subscribers 1, 37, 39 fig. 2.2, 52 table 3.2, 65 table 3.5, 130 fig. 5.3, 158, 167; in Shanghai 136, 140 fig. 6.1, 144, 148, 149
  popular organizing using 176
  telecommunications equipment for 129–31, 133–4
  text messages of 171, 176
  see also China Mobile; China Telecom; China Unicom; third generation mobile phones
monopoly:
  in China's Internet service 92
  in China's telecommunications industry 42, 46, 47, 50, 52, 59, 61, 62, 69, 76–7, 147–8
  in developing countries' telecommunications 22, 23, 24
  in France's telecommunications industry 44
  in Great Britain's telecommunications industry 21, 44
  in India's telecommunications industry 82
  in Internet provision 82
  in Japan's telecommunications industry 23, 24, 44–5
  in Mexico's telecommunications industry 22

241

Index

monopoly (*cont.*)
   in Qing dynasty telecommunications 27, 28
   in South Korea's telecommunications industry 24
   as telecommunications industry model 18–26, 45
   in the US telecommunications industry 20–1
   see also competition; competition in China
Moore, Thomas 11, 12
Motorola 134
MPT, *see* Ministry of Posts and Telecommunications (China)
Mueller, Milton 12
MultiMedia/169 85, 92
Mulvenon, James 14
municipal government, *see* local government; local government in China; Shanghai; *and underother specific cities*
Murdoch, Rupert 104

Nanjing:
   telecommunications equipment manufacturing in 125
   telephones in 28
National People's Congress 62
National Science Foundation 80, 81
national security:
   in China's telecommunications industry 46, 58, 105, 128, 142
   and telecommunications development 25
Nationalist government, *see* Republican-era government
nationalization:
   in China 10, 27, 31, 139
   in France 21
   in Great Britain 21
NEC, *see* Nippon Electric Company
NetChina 91, 92
Netcom, *see* China Netcom (old) and China Netcom (new)
Netease 101–2, 103
News Corporation 86, 104
Nextel 71
Nie, Norman 107–9
Nigeria:
   telecommunications industry in 22
Ningxia:
   Internet penetration in 172, 175
Nippon Electric Company 24, 121
   equipment sales of to China 117
   joint venture of in China 120, 121, 141
   see also Tianjin NEC

Nippon Telegraph and Telephone Corporation 24
   as China Unicom partner 56 table 3.4, 64
   Internet service of 82
   regulation of 44–5, 194n40
Noam, Eli 45
Nokia 134, 180
NSFNET 80
NTT, *see* Nippon Telegraph and Telephone Corporation
NTT DoCoMo 45, 82

Oi, Jean 138
Oksenberg, Michel 11
Organization for Economic Cooperation and Development 161

Pack, Howard 9, 188
paging services:
   of China Unicom 67, 71
Paktel 74
Pearson, Margaret 110, 117–18, 207n95
People's Bank of China 53
People's Construction Bank of China 121
People's Liberation Army 122
   Internet network of 86, 87
   investment of in Great Dragon 123, 209n47
   mobile phone network of 199n17
   telephones for 164
personal handy system:
   as telecommunications standard in China 68
Philippines:
   Internet usage costs in 94
   Internet users in 90 table 4.2
   telecommunications costs in 76 table 3.6
PHS, *see* personal handy system
Pitt, Douglas 13
plan rationality 4
pornography:
   in China's Internet 95, 96, 98, 99, 109
Posts and Telecommunications Administration:
   in China 30
   in telecommunications development model 25
private ownership in China:
   advocacy of for China's telecommunications sector 155, 156, 157, 168
   case of Huawei 126–9, 209n65
   of Internet content providers 95, 101–4, 105, 111, 184

Index

of Internet service providers 90, 92, 93, 111
  prohibition of for China's telecommunications services 45, 46, 142, 155
  in Shanghai's Internet cafés 153–4
  in Shanghai's telecommunications industry 156
  of telecommunications equipment companies 113, 127
privatization:
  of France's telecommunications 21, 44
  of Great Britain's telecommunications 21, 44
  of India's Internet network 82
  of India's telecommunications 23
  of Japan's telecommunications 23, 24, 44–5
  of Kenya's telecommunications 22
  of Mexico's telecommunications 22
  of South Korea's telecommunications 24
  of US' Internet network 80, 81
Prodigy 81
propaganda:
  spread of in China 30
provincial telecommunications administrations:
  cooperation of with Huawei 128–9, 187
  Internet service of 85, 89–90, 91 fig. 4.3, 92, 93, 102, 104
  role of in telecommunications equipment manufacture 114, 116, 120
  in Shanghai 142–3, 146, 148–9, 150
  see also Beijing Telecom, Shanghai Telecom
PTCL 74
PTIC: see China Posts and Telecommunications Industry Corporation
power-line Internet service:
  as solution for China's digital divide 180–1
Pudong:
  AT&T in 150, 151, 152
PUnet 83
Putian:
  creation of 133
  production by 133
  see also China Posts and Telecommunications Industry Corporation

Qing dynasty:
  development of telecommunications in 26–8, 139
Qinghai Province:
  mobile phones in 166 table 7.2
  per capita income levels in 172

qq.com 102 table 4.4
Qu, Weizhi 62
Qualcomm 124

Railcom, see China Railcom
Ren, Zhengfei 126–8
  founding of Huawei by 126–7
Republican-era government 32
  development of telecommunications under 28–9, 41, 113, 195n62
research and development:
  spending for by Chinese telecommunications companies 125, 127, 128, 133
Röller, Lars-Hendrick 160
Rural Electrification Administration 20, 176
rural telecommunications 45
  in China, see rural telecommunications in China
  in the US 20, 160
  see also digital divide in China, rural telecommunications in China, urban telecommunications
rural telecommunications in China 30, 33, 47, 115
  connection of villages with telephones in 165–7, 180
  government policies to improve 162–7, 176–7, 179–82
  history of 162–7
  Internet services for 158, 159, 168, 169–73, 174–6, 178–9, 180–2
  number of telephones in 163, 164, 165–6
  revenue losses in 167–8, 169, 177, 181
  telecommunications equipment sales for 125, 128, 133
  universal service fund for 176, 177, 180
  see also digital divide in China, urban telecommunications
Russia:
  Huawei investment in 128

Saggi, Kamal 9, 188
SAIL, see Shanghai Alliance Investment Company Ltd.
Samsung Mobile Communications Company 130
SARFT, see State Administration for Radio, Film, and Television:
Satcom, see China Satellite Communications
satellite dishes in China 100
satellite technology:
  for China's rural regions 179–80
SBC Communications 56

243

Index

Segal, Adam 11, 12, 17, 138, 145
Shaanxi Province:
  telecommunications development in 168
Shanda Interactive 103
Shandong Province:
  Internet education in 179
  telecommunications development in 168
Shanghai:
  China Unicom in 50, 52, 53, 54, 57, 66, 67, 147–9, 200n46
  Eastday.com in 137, 152–5
  economic growth in 135, 212n18
  fixed-line phone costs in 143, 144
  foreign investment role of in 136, 141–2, 143–4, 147–52, 155, 156, 157
  industrial policy in 136, 138, 142–6, 152, 153–4, 155, 156, 157, 185–6; in foreign-invested projects 147, 148, 149, 151
  Internet cafés in 96, 153–4
  Internet development in 86, 89–90, 93, 102, 107, 136–7, 140 fig. 6.1, 144–5, 146, 151, 152–5, 158, 169
  Internet penetration in 170
  Internet users in 136, 140 fig. 6.1, 144
  investment of in telecommunications 145
  media control of in 152, 153, 154
  mobile phone costs in 144
  mobile phones in 51, 136, 140 fig. 6.1, 144, 148–9, 166 table 7.2, 167
  municipal corporations in 136, 141 fig. 6.2, 145–6, 147, 148, 150, 151, 153–5, 156–7, 213–14n67, 214n68
  per capita income levels in 135, 143, 144, 172
  telecommunications in 136–7, 139–57, 158, 162, 168
  telecommunications equipment manufacture in 113, 114, 118–20, 121, 125, 146–7
  telecommunications equipment purchases in 143
  telecommunications history in 139–47
  telecommunications leading group in 143
  telegraph lines in 26, 27
  telephones (fixed-line) in 28, 29, 139, 140
  see also American Telephone and Telegraph Company; Eastday.com; Jiang, Mianheng; Shanghai Bell; Shanghai Unicom
Shanghai Alliance Investment Company Ltd.
  role of in Shanghai's telecommunications development 141 fig. 6.2, 145, 146
Shanghai Bell:
  controlling acquisition of by Alcatel 131–2
  cost of switches from 123
  creation of 118
  intellectual property and 121, 126, 131–2
  market share of 118, 128
  mobile phone equipment manufacture of 130–2
  production of switching systems by 118–20, 121, 125, 131, 142, 143
  profit of 118
  see also Alcatel
Shanghai Belling 119
Shanghai Cable Network
  Internet services of 146, 153
Shanghai Infoport 145
Shanghai Information Investment Corporation:
  and AT&T 146, 151
  and Eastday.com 153
  role of in Shanghai's telecommunications development 141 fig. 6.2, 146, 153
Shanghai Mobile:
  reorganization of 144
Shanghai Posts and Telecommunications Administration 142–4, 146, 147
  competition of with Shanghai Unicom 148–9
  joint venture of with AT&T 150
Shanghai Science and Technology Investment Corporation
  as China Unicom shareholder 48 table 3.1, 55 table 3.3, 56 table 3.4
  creation of 146
  role of in Shanghai's telecommunications development 141 fig. 6.2, 146
  as Shanghai Unicom partner 148
Shanghai Symphony Telecom Corporation 151
Shanghai Telecom:
  creation of 142–3
  Internet service of 153
  joint venture of with AT&T 151–2, 156
  reorganization of 144
Shanghai Trust and Investment Company 54
Shanghai Unicom 50, 52, 53, 54, 57, 66, 67, 144, 147–9, 200n46
  creation of 148
  growth of 148–9
  joint ventures in 148–9, 156

244

Index

reorganization of 148, 149
Shanghai Volkswagen 11
Shanxi:
   China Unicom in 56–7
Shen, Xiaobai 145
Sheng Xuanhuai 27, 28
Shenzhen:
   local government power of 138
   municipal corporations in 145
   telecommunications development in 158
   telecommunications equipment manufacturing in 125, 127
Shi, Tao 97–8, 100
Sichuan Province:
   telecommunications development in 168
   telecommunications equipment sales in 128
Siemens:
   joint venture of in China 119, 120, 123, 141
   mobile phone standards of in China 75
   reorganization of BISC venture by 132
   research and development spending by 133
   see also Beijing International Switching Company
SII, *see* Shanghai Information Investment Corporation
Sina.com 101–2, 103
Singapore:
   growth in 8
   industrial policy in 5, 9
   Internet usage costs in 94
   Internet users in 90 table 4.2
   telecommunications costs in 76 table 3.6
Singapore Telecom:
   as China Unicom partner 56 table 3.4, 57, 64
SK Telecom:
   joint venture of with China Unicom 202n131
Slim, Carlos 22
Softbank 45
Sohu.com 101–2
Solinger, Dorothy 10–11
Sommers, Paul 137
South Korea, *see* Korea, South
Southern Pacific Railroad Company 201–2n110
Southwestern Bell 22
Soviet Union 113, 114, 139
Sprint 53, 64, 81, 85
SSTIC, *see* Shanghai Science and Technology Investment Corporation

standards:
   for mobile phone services 67, 68, 75, 188–9
   in telecommunications equipment 45, 120
State Administration for Radio, Film, and Television:
   Bureau of in Shanghai 141 fig. 6.2, 146
   creation of 62
   investment of in China Netcom (old) 86
   policies of 94
State Commission for Restructuring the Economy 46–7
State Council 31
   'Directive 56' 37, 120, 141, 150
   'Directive 165' 50
   'Directive 178' 37, 48, 58
   Information Leading Group of 86
   policies of 36, 50, 200n69; toward China Unicom 37, 48, 49, 53, 54, 58, 60, 76–7; on China's Internet 83, 85–6, 95–6; on China's telecommunications equipment sector 37, 117, 120, 129, 141, 150
   role of in telecommunications 50, 77
State Development Planning Commission 66, 177
State Economic and Trade Commission 58, 60
State Economic Commission 31
State Education Commission 83
State-Owned Assets Supervision and Administration Commission 70–1
State Planning Commission:
   planning of for telecommunications equipment manufacturing 115
   policies of toward China Unicom 53–4, 58
   regulation of telecommunications by 31, 36
State Radio Regulatory Committee 68
State System Restructuring Commission 58
Steinfeld, Edward 11, 12, 15
Stet 56 table 3.4
Sumitomo 120
Sun Microsystems 85
Sun, Yafang 127
switching systems, *see* telecommunications equipment manufacture in China: switching systems in, and telecommunications equipment manufacture in China: switching systems pricing in

Taiwan 109
   growth in 8
   industrial policy in 5, 6, 7, 9
   telegraph lines to 27

Index

Tan, Zixiang (Alex) 12, 13
Tang, Shaoyi 28
Taobao 103
Taubman, Geoffry 13
TD-SCDMA 75
technocrats:
  in China's telecommunications leadership 36, 40–1
technology transfer:
  in China Unicom case 54–5
  in China's telecommunications equipment manufacture 112, 113, 117–18, 119, 121, 127, 131–2, 150
  in Shanghai-AT&T cooperation 150, 151
  in Shanghai Unicom case 149
  as tool for developing countries 5, 156, 189
TELECOMM 22
telecommunications:
  benefits of for municipalities 137
  costs of service in selected nations 76 table 3.6
  development of: in China, *see* telecommunications in China; in France 21; in Great Britain 21; in Japan 23–4; in the US 19–21, 25–6, 160
  regulatory models of 18–26; in developing countries 21–5
  relation of to GDP growth 160, 214n2
  see also telecommunications in China
telecommunications in China
  competition in, *see* competition in China
  digital divide of 94, 107, 158–9, 161–82
  foreign investment in, *see* foreign investment in China
  history of, *see* China: telecommunications history of
  Internet development in, *see* Internet in China
  law for 54, 62
  mobile phones in, *see* mobile phones in China
  policies for, *see* industrial policy in China
  in Shanghai 136–7, 139–57
  standards for, *see* standards
  studies of 12–14
  telecommunications equipment for, *see* telecommunications equipment manufacture in China
  total fixed-asset government investment in 37, 38 table 2.1
  see also digital divide in China; foreign investment in China; industrial policy in China; Internet in China; joint ventures; mobile phones in China; rural telecommunications in China; Shanghai; telecommunications equipment manufacture in China; telephones in China (fixed-line); urban telecommunications; *and under specific Chinese government ministries and telecommunications companies*
telecommunications equipment manufacture:
  in China, *see* telecommunications equipment manufacture in China
  in India 23
  in Japan 23–4
  in Kenya 22
  in the US 20
  see also telecommunications equipment manufacture in China
telecommunications equipment manufacture in China 31, 37, 47, 112–13, 198n12
  government policies of 112, 121–2, 123, 124, 126, 134
  history of 113–21, 129–34
  import duties and 117
  import substitution and 112, 116–17, 119, 120–1, 128, 130, 134
  intellectual property rights protection and 121, 126, 131–2
  joint ventures in 112, 117–20, 121–2, 125, 126, 129–32, 133, 134, 155
  localization of 119, 121
  private companies in 113
  production capacity of 118, 119 table 5.1
  research and development spending for 125, 127, 128, 133
  switching systems in 114, 115, 117, 118–24, 125–26, 127, 128, 129–31, 132, 134
  switching systems prices in 121, 123–4
  technology transfer in 112, 113, 117–18, 119, 121, 127, 131–2
  see also Alcatel; American Telephone and Telegraph Company; Beijing International Switching Company; Great Dragon Group; Huawei Technologies; Shanghai Bell; Siemens; Tianjin NEC; Zhongxing Telecommunications Equipment Company
teledensity:
  in Argentina 22
  in China 30, 33, 36, 37, 38, 39 table 2.2, 158, 197n104

246

Index

in India 22, 23, 39 table 2.2
in Japan 24, 25, 39 table 2.2
relation of to GDP growth 160
in selected nations 39 table 2.2
in Shanghai 136, 144
in South Korea 25, 39 table 2.2
in the US 24
in Western Europe 24
telegraph network:
  in China 26–8, 29, 33
  in France 21
  in Great Britain 21
  in Japan 23
  in the US 19
telephones in China (fixed-line):
  costs and installation fees for 38, 76 table 3.6, 143, 144, 164, 165, 168, 179
  number of subscribers, 1, 28, 29, 30, 31 fig. 2.1, 37, 39 fig. 2.2, 115, 122 fig. 5.2, 140, 143, 163, 164, 165; in rural areas 163 table 7.1; in Shanghai 139, 140; in urban areas 163 table 7.1
  in Qing dynasty 27, 28
  in Republican era 29
  see also mobile phones in China, telecommunications in China
Telesystem International:
  as China Unicom partner 56 table 3.4, 57, 64
television:
  in China 30, 105
  see also Internet in China: cable television used for
TELMEX 21
Telstra 202n131
Tengtu International Corporation 179
Texas Instruments 128
Thailand:
  Internet usage costs in 94
  Internet users in 90 table 4.2
  telecommunications costs in 76 table 3.6
  telephones and teledensity in 39 table 2.2
third generation mobile phones:
  for China Satellite Communications 71
  licenses for 73, 75, 77
  standards for 67, 75, 188–9
3G, see third generation mobile phones
'Three 90 percents':
  telecommunications policy of 36, 38, 140
  telecommunications policy of in Shanghai 143

Thun, Eric 17, 138, 145
Tian, Gang 138, 145
Tiananmen demonstrations of 1989 47, 97, 107, 175
Tianjin 60
  China Unicom in 50, 53, 61, 63, 67, 148
  Internet penetration in 170
  mobile phones in 166 table 7.2
  telecommunications equipment manufacture in 114
  telegraph lines in 27
  telephones in 29
Tianjin NEC:
  creation of 120
  problems of 121
Tianjin Telecommunications Authority 120
Tianjin Zhonghuan Computer 120
Tibet:
  CERNET in 178
  illiteracy in 174
  Internet development in 107
  Internet penetration in 170
  mobile phones in 166 table 7.2, 167
  per capita income levels in 172 table 7.4, 217n58
total factor productivity 5–6, 7, 8
  definition of 190n15
Transparency International 187
'triple play':
  communications services of in Shanghai 146
TUnet 83
'Two Six-Point Targets' 36

Unicom, see China Unicom
Unicom Investor Group 64
Uninet 86, 87, 88 table 4.1
United Kingdom, see Great Britain
United National Educational, Scientific, and Cultural Organization 179
United States 109
  development of Internet in 80–1
  Internet usage patterns in 107–9
  Internet users in 90 table 4.2
  investment of in China 118
  municipal telecommunications in 137
  power-line Internet in 180
  telecommunications development in 19–21, 25–6, 160
  telephones and teledensity in 39 table 2.2
  universal service fund in 176–7
  see also American Telephone and Telegraph Company; and under other specific American companies

247

Index

universal service:
  fund for in China 176, 177, 180
  fund for in the US 176–7
  in telecommunications development model 25, 26, 45
urban telecommunications:
  in China's Internet 107, 168, 169, 170, 171, 172, 173, 178–9
  in China's telephone network 29, 115, 158, 162, 163, 164, 165, 166, 167, 168, 173
  number of telephones in 163, 164, 165
  in telecommunications development models 25–6
  see also digital divide; digital divide in China; rural telecommunications in China; *and under specific Chinese cities*

Vallat, Maurice 121
Videsh Sanchar Nigam Ltd. 82
Vietnam:
  Internet users in 90 table 4.2
  Internet usage costs in 94 table 4.3
  telecommunications costs in 76 table 3.6
  telephones and teledensity in 39 table 2.2
village telecommunications, *see* rural telecommunications in China
Vodaphone 44
Vogel, Ezra 5
voice over Internet protocol 66
Volkswagen 212n16

Walcott, Susan 137
Walder, Andrew 110
Wang, Jianzhou 63, 74, 77
Wang, Xiangdong 14
Wang, Xiaochu 77
Wang, Zigang 35
Waverman, Leonard 160
W-CDMA 75
Wen, Minsheng 35, 116
Western Electric Manufacturing Company 19, 20, 24, 44
Western Union Telegraph company 19, 20
Wheeler, Susan 137
White, Lynn 152
Wignaraja, Ganeshan 7
wireless shortwave 30
Wong, Chee Kong 13
World Bank:
  loans of for China 71, 117
  policies of toward China 47, 198n11
  report of on industrial policy 5, 7, 190n2

World Trade Organization:
  China's entry to 2
  effect of on China's Internet 80, 104, 105
  effect of on China's telecommunications industry 72, 73–4, 78, 151, 152, 156, 188
  Reference Paper of 74
  rules of on intellectual property rights protection 132
WTO, *see* World Trade Organization
Wu, Bangguo 58
Wu, Jiangxing 122, 123, 124
Wu, Jichuan 49, 60, 61, 62
  Internet policies of 85, 86, 100–1
  policies of toward AT&T 151
  policies of toward Shanghai Bell 131
  rural telecommunications policies of 165
  telecommunications policies of 40–1, 63, 68
Wu, Wei 13
Wuhan:
  China Unicom in 66

'xiang xiang tong dianhua' policy 162
Xiaolingtong:
  mobile network for 68–9, 77, 132
Xingfa Chemicals Group 90
Xinhua News Agency 102
Xinjiang:
  Internet penetration in 172, 175
Xu, Kuangdi 151
Xu, Li 168

Yahoo:
  arrest of Shi Tao and 97–8
  Internet service of in China 100, 102 table 4.4, 104
Yan, Xu 13
Yang, Changji 60
Yang, Guobin 109
Yang, Taifang 35–6, 40, 49, 116, 118
Yang, Xianzu 62, 63
Yunnan Province:
  illiteracy in 174
  Internet development in 107, 158
  telegraph lines to 27
Yunxing Electronic Trading Company 125

Zhang, Jianhong 98
Zhang, Ligui 68
Zhang, Shuxin 90, 92, 97
Zhao, Weichen 49, 60, 61, 63

Index

Zhejiang Province:
   Internet penetration in 170 table 7.3
   Internet service in 90, 169
   mobile phones in 166 table 7.2, 167
   telephones (fixed-line) in 29
zhengqi fenkai 50
Zhong Fuxiang 33, 35
Zhongxing Telecommunications Equipment Company 123, 124, 183
   creation of 124–5
   exports by 125–6, 133, 189
   market share of 126
   production of 125–6, 133, 134
   research and development spending by 125
   revenue of 133
   strategies of 125–6, 133
Zhou, Deqiang 62, 68
Zhou, Shaojie 168–9
Zhu, Entao 96
Zhu Rongji 41, 48, 60, 62, 92, 110
Zhu Xuefan 32, 33, 35
Zongli Yamen 26
Zou, Jiahua 48, 58, 61
ZTE, *see* Zhongxing Telecommunications Equipment Company munica